As issues of equity and justice move centre stage worldw ··········
a responsive, unified, intersectional environmentalism? T ··········
but Bell has assembled an excellent collection that utilizes academic, personal and
experiential perspectives to give us fresh, creative new ideas.

Julian Agyeman, Professor of Urban and Environmental
Policy and Planning, Tufts University

A well-timed, much needed holistic approach to the ways in which social inequalities such as class and disability are manifested in mainstream environmentalism.
Everyone involved in environmental campaigning should read this book!

Lucie Marks, Key Relationships Officer, Greenpeace UK

This book pulls together social movements from around the world that have been
at the frontline of fighting for climate justice. It will support the growing movement
for socially just environmental policies like the Green New Deal.

Natasha Josette, Community Organiser, Green New Deal

This book is a much needed, detailed dissection of the lack of diversity within the
environmental sector. It is thoroughly researched and sets out everything the sector
needs to know to make the change.

Mya-Rose Craig AKA Birdgirl, conservationist,
environmentalist and race activist

DIVERSITY AND INCLUSION IN ENVIRONMENTALISM

This book discusses how to develop green transitions which benefit, include and respect marginalised social groups.

Diversity and Inclusion in Environmentalism explores the challenge of taking into account issues of equity and justice in the green transformation and shows that ignoring these issues risks exacerbating the gap between the rich and the poor, the marginalised and included, and undermining widespread support for climate change mitigation. Expert contributors provide evidence and analysis in relation to the thinking and practice that has prevented us from building a broad base of people who are willing and able to take the action necessary to successfully overcome the current ecological crises. Providing examples from a wide range of marginalised and/or oppressed groups including women, disabled people, Black, Asian and Minority Ethnic (BAME) people and the lesbian, gay, bisexual, transgender, queer/questioning and others (LGBTQ+) community, the authors demonstrate how the issues and concerns of these groups are often undervalued in environmental policy-making and environmental social movements. Overall, this book supports environmental academics and practitioners to choose and campaign for effective, equitable and widely supported environmental policy, thereby enabling a smoother transition to sustainability.

This volume will be of great interest to students, scholars and practitioners of environmental justice, social and environmental policy, planning and environmental sociology.

Karen Bell is a Senior Lecturer in Human Geography at the University of the West of England, UK. She is an inter-disciplinary social scientist who has been investigating and teaching at the intersection of political, geographical and environmental studies for the last ten years.

Routledge Studies in Environmental Justice

This series is theoretically and geographically broad in scope, seeking to explore the emerging debates, controversies and practical solutions within Environmental Justice from around the globe. It offers cutting-edge perspectives at both local and global scales, engaging with topics such as climate justice, water governance, air pollution, waste management, environmental crime and the various intersections of the field with related disciplines.

The Routledge Studies in Environmental Justice series welcomes submissions that combine strong academic theory with practical applications, and as such is relevant to a global readership of students, researchers, policymakers, practitioners and activists.

Ecosocialism and Climate Justice
An Ecological Neo-Gramscian Analysis
Eve Croeser

Climate Change Justice and Global Resource Commons
Local and Global Postcolonial Political Ecologies
Shangrila Joshi

Diversity and Inclusion in Environmentalism
Edited by Karen Bell

Environmental Justice in the Anthropocene
From (Un)Just Presents to Just Futures
Edited by Stacia Ryder, Kathryn Powlen, Melinda Laituri, Stephanie A. Malin, Joshua Sbicca and Dimitris Stevis

For more information about this series, visit: www.routledge.com/Routledge-Studies-in-Environmental-Justice/book-series/EJS

DIVERSITY AND INCLUSION IN ENVIRONMENTALISM

Edited by Karen Bell

Routledge
Taylor & Francis Group

LONDON AND NEW YORK

earthscan
from Routledge

First published 2021
by Routledge
2 Park Square, Milton Park, Abingdon, Oxon OX14 4RN

and by Routledge
52 Vanderbilt Avenue, New York, NY 10017

Routledge is an imprint of the Taylor & Francis Group, an Informa business

British Library Cataloguing-in-Publication Data
A catalogue record for this book is available from the British Library

Library of Congress Cataloging-in-Publication Data
Names: Bell, Karen (Karen Frances), 1959–
Title: Diversity and inclusion in environmentalism / edited by Karen Bell.
Description: Abingdon, Oxon; New York, N.Y.: Routledge, 2021. |
Series: Routledge studies in environmental justice |
Includes bibliographical references and index.
Identifiers: LCCN 2020053861 (print) | LCCN 2020053862 (ebook) |
ISBN 9780367567309 (hardback) | ISBN 9780367567354 (paperback) |
ISBN 9781003099185 (ebook)
Subjects: LCSH: Environmentalism–Social aspects. | Environmental policy–Social aspects. | Sustainability–Social aspects. | Environmental justice.
Classification: LCC GE195.D556 2021 (print) |
LCC GE195 (ebook) | DDC 363.7008–dc23
LC record available at https://lccn.loc.gov/2020053861
LC ebook record available at https://lccn.loc.gov/2020053862

ISBN: 978-0-367-56730-9 (hbk)
ISBN: 978-0-367-56735-4 (pbk)
ISBN: 978-1-003-09918-5 (ebk)

Typeset in Bembo
by Newgen Publishing UK

For those engaged in struggles for social and environmental
justice, everywhere

CONTENTS

ACKNOWLEDGEMENTS

There are many people involved in creating a book and I wish to thank them all. Firstly, I would like to thank the chapter authors who have taken on the challenge of trying to condense information and analysis on the huge range of equalities issues that pertain to environmentalism in their specialist field. They have all worked diligently to bring this book to fruition. Among them, I want to give a special thank-you to Clara Greed, who, along with helping with reviewing, encouraged me to take the first steps to make my dream of coordinating such a book into reality. In addition, I am very grateful to the anonymous reviewers of the proposal and chapters, and to others who have helped with reviewing, including Valerie Walkerdine, Finn Mackay and Emma Foster. I would also like to thank Routledge staff for their enthusiastic response to my book proposal and their efficient handling of the publishing of the book. In particular, I would like to thank the editors, Mathew Shobbrook and Annabelle Harris. I am also deeply grateful to all the authors referenced in this book for the work they have carried out that relates to diversity, inclusion and environmentalism. Their work has informed and inspired me for many years.

I would also like to thank the staff at the University of the West of England, Bristol (UWE) which has been my academic home for the last two years. The staff and students on the BA Geography Team are a real pleasure to work with and I am constantly learning from them. Members of the wider Faculty of Environment and Technology have also been extremely helpful with their advice, practical support and friendliness, especially Lindsey McEwan and Jonathon Stadon. Most importantly, I want to thank the support staff at UWE who perform a great job at keeping everything clean, functioning, safe and organised. It would be impossible for me to do my job or write anything without them.

CONTRIBUTORS

Becky Alexis-Martin, Lecturer in Political and Cultural Geographies, Manchester Metropolitan University

Dr Becky Alexis-Martin is an award-winning pacifist activist scholar, author and photographer. Her first book, *Disarming Doomsday*, critically considered the inequalities and harms perpetuated by nuclear warfare. *Disarming Doomsday* was shortlisted for the 2020 Bread and Roses Award and was the recipient of the 2020 L.H.M. Ling Outstanding First Book Prize.

Karen Bell, Senior Lecturer in Human Geography and Environmental Justice, University of the West of England, Bristol, UK

Formerly a youth and community development worker, for 15 years Karen worked alongside disadvantaged communities to address issues such as social inequality, racism, disability discrimination, and environmental exclusion and inequity. As an academic, her work has looked at how to achieve environmental justice in a variety of political, economic and cultural settings; how to build a fair and inclusive transition to sustainability; and how to build an environmental movement that is attractive to a wider range of social groups.

Gnisha Bevan, Organiser, Black Seeds Environmental Justice Network

Gnisha co-developed and coordinates the Black Seeds Environmental Justice Network which seeks to recognise and support environmentalists of colour. She works as a consultant advisor on a range of environmental and social justice projects and is currently studying for an MSc degree in Sustainable Development in Practice at the University of the West of England, Bristol. She is an education specialist and has worked on a number of international projects, including collaborating with the Rwandan Education Board, international development partners and the voluntary sector to improve the quality of education in Rwanda.

Benjamin Bowman, Lecturer in Youth Justice, Department of Sociology, Manchester Metropolitan University
Benjamin Bowman is a Lecturer in Youth Justice and a member of the Manchester Centre for Youth Studies. He is the co-convener and treasurer of the Young People's Politics Specialist Group of the Political Studies Association. His research is on young people's everyday politics, and the opportunities for social change in contemporary democracy. Benjamin has worked extensively with young people, including young environmentalist activists, and is a contributor to the international working group on the Existential Toolkit for Climate Educators, Rachel Carson Center, Munich.

Emma Foster, Lecturer in International Politics, University of Birmingham, UK
Emma's research explores the interrelations and tensions between gender, sexuality and ecologism/environmentalism. Informed by feminist and queer theory, her interests include alternative approaches to (sustainable) development and environmental politics. She is currently researching ecofeminist critiques of technology in the context of the Anthropocene.

Deborah Fenney, Researcher, Policy Team, The King's Fund, UK
Deborah Fenney researches policy at the King's Fund, with a particular interest in inequalities in relation to disability and sustainable lifestyles; and the barriers disabled people face to accessing environmentally friendly initiatives.

Clara Greed, Emerita Professor of Inclusive Urban Planning, University of the West of England, Bristol, UK
Clara Greed is Emerita Professor of Inclusive Urban Planning and a chartered town planner. Her research interests and publications centre on urban design, 'women and planning', the built environment professions, 'sanitation and sustainability' and social infrastructure. She has authored over 160 publications and was awarded an MBE for services to urban design in 2009.

Roger Griffith, Organiser, Black Seeds Environmental Justice Network
Roger Griffith MBE is a consultant and chief executive officer (CEO) of his company Creative Connex. He is the author of *My American Odyssey: From the Windrush to the White House*. He is a social activist, lecturer at UWE Bristol, broadcaster and former CEO/chair of Ujima Radio, an award-winning community radio station. He is also a creative producer and sits on the Arts Council England South-West board. He has a passion for sharing cultural stories, global observations and insights on race, inclusivity and social inequality and has been delivering diversity training for over 30 years.

Gary Haq, Senior Research Fellow, Stockholm Environment Institute
Gary Haq is a human ecologist with over 20 years of experience in undertaking research on a wide range of environmental policy issues. He is a Senior Research Fellow at the Stockholm Environment Institute based in the Department of

Environment and Geography at the University of York (UK). He was a visiting research scientist at the Directorate for Energy, Transport and Climate, European Commission's Joint Research Centre (Italy), providing technical support for the development of European environmental and transport policy. Gary has authored numerous reports and academic papers on transport, air pollution, climate change, behaviour and lifestyle. He has written for *The Guardian*, *Yorkshire Post*, *The Independent* and *The Conversation* on a range of environmental issues.

Aleksandra Kosanic, Lecturer, Liverpool John Moores University, UK

Aleksandra Kosanic is a physical geographer interested in interdisciplinary approaches to explore climate change impacts on multiple ecosystem services and society. She looks at how the impacts will affect particular communities (e.g. indigenous, disabled, elderly and youth). She has written about how climate change and the loss of ecosystem services will affect the world's disabled populations disproportionately by exacerbating inequalities and increasing marginalisation. They may experience limited access to knowledge, resources and services, which may prevent them from effectively responding to climate change, and disabled populations may also prove more vulnerable to extreme climate events.

Harriet Larrington-Spencer, Researcher, Healthy Active Cities, University of Salford, UK

Harriet's research interests centralise around environmental sustainability, with an emphasis on everyday urban mobility and inclusive active travel. She is particularly interested in disabled environmentalism and is currently exploring intersections of feminist theory and crip theory to inform this work.

Lucie Middlemiss, Associate Professor in Sustainability, Leeds University, UK

Lucie Middlemiss is interested in the boundary between social and environmental issues, with a focus on how and whether people are able to participate in environmental policy and practice. She has particular interests in how people's relationships and their membership of broader social categories and allegiances shape their actions in everyday life. Her subject interests include sustainable consumption, sustainable communities, environmental justice and energy poverty.

Silpa Satheesh, Assistant Professor, School of Development, Azim Premji University, India

Silpa's research focuses on the relationship between labour and environmental movements, environment and development, postcolonial social movements, political economy and ethnography. As a social movement scholar, she works in close collaboration with grassroots environmental movements and environmental organisations in Kerala, India. Silpa writes about her research projects in academic as well as popular outlets in both English and Malayalam. She recently completed her PhD in Sociology from the University of South Florida.

1

DIVERSITY AND INCLUSION IN ENVIRONMENTALISM

Karen Bell

Introduction

This book grew out of the belief that everyone has a place in the environmental movement and that environmentalism has an essential place in the equalities movements. It aims to build and develop bridges between the equalities and
environmental communities. To do this, the authors encourage environmentalism
to become an inclusive and diverse coalition, and advocate a widening of the
equalities agenda to include the relevant sustainability issues. It is our hope that
reading this book could support the process of generating these understandings
and connections, and creating a movement that unites environmental and social
justice. This movement would be effective and relevant, addressing the needs and
perspectives of all and valuing all voices.

At a time of supposed 'culture wars' (Hunter, 1991) where, it is argued, society
has become increasingly polarised, sometimes around identity issues, it seems
important to stress that this book is about facilitating solidarity and not creating
divisions. The divisions are already there but we cannot eradicate them unless we
are honest about the barriers to creating unity. These barriers include attitudes,
beliefs, policies, practices and cultures. But this is not solely about oppression – the
forthcoming chapters often point out how different diverse groups are actually
or potentially active contributors to environmentalism. In this context, the book
authors have endeavoured to explain the barriers to engaging with mainstream
environmentalism and offer suggestions as to how to overcome these hindrances.
In addition to having academic expertise, the authors almost all come from the
backgrounds they are writing about. They write from personal experience, as well
as knowledge based on their academic research and/or community practice. It is
impossible to be comprehensive in the one chapter dedicated to each social demographic. However, each chapter highlights some of the main issues that need to be

addressed and provides insights and resources which will hopefully encourage and enable further reading and investigation.

Language is, of course, important in discussions of diversity and inclusion. The separate chapters discuss the rationale for the specific language used in relation to the group focussed on. When discussing, collectively, the eight socio-economic groups addressed in this book, that is, women, disabled people etc., we use general terms such as 'marginalised people/groups/communities', 'disadvantaged people/groups/communities', 'equalities people/groups/communities' and 'oppressed people/groups/communities'. While these terms may not exactly apply to all of the eight demographic groups highlighted in the chapters, they are used here as short-hand terms.

From different positions, according to each author's positionality and expertise, this book will identify, explain and make recommendations on how to overcome the barriers to creating a diverse and inclusive environmental movement, as well as an environmentally focussed equalities movement. It will be of use to environmental activists, environmental practitioners, urban and rural planners, environmental policymakers, equalities activists, equalities practitioners and equalities policymakers. The first section of this chapter discusses the rationale for the book. This is followed by a section on definitions so as to clarify the scope and focus. The final section of the chapter describes the structure and contents of the book.

Why do we need inclusion and diversity in environmentalism?

The main reason that we need greater inclusion and diversity in environmentalism is that we need to bring as many on board as possible to the cause of transitioning to sustainability. This will help to ensure that sustainability can be achieved rapidly, effectively and equitably. We are facing multiple environmental crises and, therefore, we need to address this challenge in a timely and just manner. In 2015, the Stockholm Resilience Centre published a report showing that, as the result of the overuse and misuse of resources, we have either crossed or are about to cross, nine earth system 'planetary boundaries' beyond which there will be irreversible impacts such that the survival of humanity will be threatened (Steffen et al., 2015). We have already overstepped some of these boundaries, with climate change but one of the nine boundaries that we are currently transgressing. It is already possible to see the results of this in the form of more intense floods, hurricanes, heat-waves, droughts and famines. Similarly, the Intergovernmental Panel on Climate Change (IPCC, 2018) stated that we must reduce carbon emissions to net zero by 2050 in order to have a reasonable chance of limiting global warming to the critical 1.5°C. They warned then that we may have only 12 years to avoid irreversible negative consequences resulting from overstepping this boundary (IPCC, 2018). The report advocated rapid, far-reaching and unprecedented societal transformation to avert the worst scenarios.

Following this, the first Global Assessment of the Intergovernmental Science-Policy Platform on Biodiversity and Ecosystem Services report (IPBES, 2019) was published, which found widespread, accelerating and unprecedented declines in the Earth's biodiversity. It warned that by jeopardising the health of ecosystems, we are eroding the foundations of our economies, livelihoods, food security, health and quality of life (IPBES, 2019). Again, the report stated that it is not too late to make a difference, but only if we start now at every level from local to global. This year, another concerning report was published, which reinforced this message. The latest of the regular Living Planet Reports of the World Wide Fund for Nature (WWF, 2020) stated that our planet's wildlife populations have now plummeted by 68% since 1970.

Just these four assessments alone indicate the urgency and seriousness of the situation and the necessity to undertake a rapid and effective transition to sustainability. They also show that technical innovation will not be enough. Structural and institutional change is needed as well as a wide-scale transformation in the beliefs, attitudes and behaviour of individuals. These are just a few of the issues among very many that we now face, indicating the urgency to bring as many on board as possible to the task of transitioning to sustainability. A large mass movement of environmentalists can pressurise politicians, the media, scientists and businesses to focus their efforts on addressing the problems described. Though many are now coming around to supporting environmentalism, given the urgency and gravity of the situation, we would expect to see more people pressing for this change. Therefore, work to enable greater involvement is required, and this includes some self-reflection on the part of environmentalists.

Why environmental campaigns need to change

This is a crucial time to look at how bridges can be developed between environmentalists and equalities communities so as to ensure that environmental issues are adequately addressed in ways that are attuned to social and cultural difference. As this book describes, one reason that environmentalism has been limited in terms of membership and impact is that it has not been adequately diverse and inclusive. It is not alone in failing to be as inclusive as it could be – many contemporary social movements are divided along lines of race, sexuality, ethnicity, and class, among other issues (Echols, 1989; Gitlin, 1995; Tarrow, 1998; Snow and McAdam, 2000). However, given the urgency of the situation, it is particularly important to address this issue within environmentalism. If the environmental movement can be more inclusive, it can improve its influence on policy and society.

Another reason that it is essential to include equalities considerations in environmentalism is that environmental degradation and its impact on human well-being is not distributed evenly. It tends to reflect underlying patterns of oppression (Bullard, 1990; Bryant and Mohai, 1992; Morello-Frosch et al., 2002; Pellow and Park, 2003; Prakash, 2007). For example, it is now well documented that, in general, minority ethnic and working-class people in the United Kingdom (UK) and the United States (US) tend to live nearer to polluting facilities (e.g. Heiman, 1996; United Church of Christ 1987, 2007; FoE, 2001; Walker et al., 2003). They are also more likely to be exposed to environmental contaminants, in general (Cutter and Solecki, 1996; Morello-Frosch et al., 2002; Morley, 2006; Milojevic et al., 2017). The environmental justice movement has drawn attention to these disparate burdens (Taylor, 1993; Sze and London, 2008; Haluza-Delay et al., 2009).

Environmental justice academics and analysts have also highlighted the White[1], middle-class, male, heteronormative and disablist nature of mainstream environmentalism in the Global North (e.g. Taylor, 1993; Sze and London, 2008; Haluza-Delay et al., 2009; Bell, 2016; Feliz Brueck and McNeill, 2020). This critique has been applied to both its membership and campaign focus (e.g. Finney, 2004; Harper, 2009; Palamar, 2008; Taylor, 1993, 2000). Some might argue that we should not focus on identities which emphasise difference but, rather, on commonalities in order to build a movement. However, existing scholarship on social movements that attribute success to shared identities does not take into account relations of domination among its activists. Domination occurs, as Young (2000, p. 32) observed, where '… other persons or groups can determine without reciprocation the conditions of their actions'. When some social groups dominate others, there can be greater levels of misunderstanding and conflict. Such domination obstructs the development of shared identities (Weldon, 2006), for example, as environmentalists. Historically marginalised groups, even if nominally included, often perceive more privileged groups in social movements to be dominating in decision-making. At the same time, those from the more privileged groups have a tendency to dismiss the issues raised by the non-dominant groups, compounding mistrust between groups and possibly even leading to the decline of the movement (Williams, 1998; Mansbridge, 1999).

If marginalised groups can have a stronger voice in the environmental movement, they can dedicate their time to working on environmentalism and not struggling to be heard. This is particularly important for those who experience simultaneous oppressions (e.g. a disabled, young, lesbian), who are even less likely to be heard. Power imbalances among social movement members distort and impede communication (Habermas, 1987; Young, 2000). In order to make a difference, it is important to construct coalitions across these divisions. Diversity is essential because the presence of members of marginalised groups helps to ensure that the tactics, messages and solutions proposed reflect the interests of the marginalised group. At the same time, the visible participation of members of marginalised groups in actions and deliberations increases the trust in the movement by other members of the oppressed group. The framing of environmental problems and the solutions and policies proposed depend very much on who is involved in the conversation. If we don't have a diversity of people involved, we cannot develop solutions that are fair, with popular buy-in, and which will work for everybody.

However, marginalised groups may want to organise themselves separately, as some of those discussed in this book have. In this way, they have the opportunity to develop and voice their distinctive perspectives. It has been argued that when dominated groups have a separate discussion among themselves, they are better able to counter their marginalisation in the broader public sphere (Fraser, 1992; Fraser, 1995b; Young, 2000). These experiences can serve as training for addressing inequities, providing a space where marginalised groups can develop new concepts and ideas (Fraser, 1992).

It is also necessary to consider that environmental movements exist within a context of structural inequality. The hierarchies, divisions and conflicts that permeate society also affect social movements. If we could eliminate the underlying social inequality, diversity and inclusion practices might be less important. But, for now, we are faced with deep inequality and, therefore, must consider how we can operate as environmentalists in the midst of this situation.

What is 'diversity' and 'inclusion'?

Numerous definitions of diversity and inclusion can be found in the literature. Diversity tends to focus on who is involved or engaged, while inclusion tends to be more interested in how this is done. With regard to who is involved, some definitions focus on groups protected by national legislation, which, in the UK, are those covered by the Equality Act 2010. This aims to protect people from being discriminated against on the basis of nine characteristics, that is, sex; race; disability; age; sexual orientation; marital or civil partnership status; pregnancy and maternity; religion or belief; and gender reassignment. Hence, these are known as 'protected characteristics'. It is notable that social class and socio-economic status are not included as protected characteristics in the UK. Other considerations and definitions of who needs to be involved are wider. For example, Cox (1993, p. 5) describes inclusion as 'the representation, in one social system, of people with distinctly different group affiliations of cultural significance'. This definition could include class and other identities that are not mentioned in the national legislation of a particular country. It should also be noted that diversity can include visible and invisible difference.

Inclusion differs from diversity in focusing, not only on the compositional mix of people, but also on their incorporation into the organisation or movement. Inclusion can be considered as 'the degree to which those affected by [a decision] have been included in the decision making processes and have had the opportunity to influence the outcomes' (Young, 2000, p. 5). A more recent definition would be that of Ferdman (2017, p. 235), which states that 'In inclusive organizations and societies, people of all identities and many styles can be fully themselves while also contributing to the larger collective, as valued and full members'. Overall, then, inclusion is about freedom of expression and being part of decision-making and equally valued.

Benefits of considering diversity and inclusion

The main benefits that have been proposed for diversity and inclusion are, firstly, that these are steps towards social justice and, therefore, ethically desirable; and, secondly, that people with a variety of backgrounds and characteristics bring advantages to an organisation in terms of creativity and problem-solving (Ferdman,

2014; Hossain et al., 2019). Without a commitment to, and action on, inclusion and diversity in environmentalism, organisations and movements will not attract and/or retain the full range of humanity. In relation to addressing environmental issues, we need to bring in a variety of perspectives and experiences so as to find appropriate and publicly acceptable solutions.

However, it has been noted that diversity alone does not always bring immediate benefits to organisations (Mannix and Neale, 2005; Jackson and Joshi, 2011), in the sense that it can increase conflicts and even lower performance. What is needed are helpful approaches to inclusion that enable the potential benefits of diversity (Mor Barak et al., 2016). Downey et al. (2014) found that diversity practices only engender trust when inclusion practices are present. It is not enough to convene a diverse group, the members of the group or movement need to be treated in a way that is inclusive.

Shore et al. (2018) outline a number of themes that indicate effective 'inclusion'. Firstly, members of marginalised and disadvantaged groups need to feel safe, psychologically and physically, to share different opinions and views from others (see also Carmeli et al., 2010; Hirak et al., 2012). This could be from comfortably expressing a minority view to pointing out discrimination. Secondly, people from all backgrounds should feel like they belong, can consider themselves as an insider and can access important information and resources (see also Mor Barak and Cherin, 1998; Shore et al., 2011). Furthermore, inclusion requires that all need to feel respected and valued, as individuals and as members of an equalities or identity group (see also Nishii, 2013; Tang et al., 2015). If remarks are made about the group, it can wound as deeply as remarks being made about an individual. Moreover, everyone needs to be able to influence the relevant decision-making. People need to be listened to and taken seriously. Underpinning all of these, there needs to be recognition of diversity in the sense that not everyone is the same or has the same needs.

Moreover, a key benefit for environmentalism from greater diversity and inclusion is that the range of environmental actions proposed will be achievable by a wider range of people, both financially and culturally. Currently some actions, such as changing consumption patterns, civil disobedience, lobbying members of parliament and even participating in letter-writing campaigns, may not be universally accessible.

Limits of diversity and inclusion

It has been argued that since diversity and inclusion do not require structural change or major transformation, they are going to be extremely limited in their impacts (e.g. Walcott, 2019). They merely enable inclusion into the current structures, without changing those structures significantly. From this perspective, diversity and inclusion can arguably be seen as centring the dominant groups, whether White, male, middle-class, non-disabled, heterosexual, etc. It is clear that we need to consider what we are diverging from and what are we being included in. Diversity and inclusion could, otherwise, seem to imply that those who are not of the dominant group aspire to be part of institutions or programmes that the dominant group owns or controls, without any fundamental change on the part of that dominant group or its institutions and projects.

'Inclusion' has emerged alongside the concept of 'social inclusion'. Ronald Labonte, among others, has discussed some of the problems with the idea of social inclusion, including that its focus on specific problems avoids confronting the structural reasons for exclusion and the advantages that such exclusion can have for the dominant groups in a society. He asks, 'how does one go about including individuals and groups in a set of structured social relationships responsible for excluding them in the first place?' (Labonte 2004, p. 117). Inclusion policies and practices seem to assume that excluded populations desire to be included in what currently exists, rather than transforming this status quo.

Clarifying this tension, Fraser (1995a) outlines two distinct remedies for social injustice: affirmative and transformative. Affirmative remedies are intended to correct inequitable outcomes without altering the structures that initiate, underpin and maintain the inequity. Transformative remedies aim to dismantle the structures and systems that drive inequity in order to address the injustice at the root. This book addresses both of these. We are asking, 'who is excluded, why and with what consequences?' In general, the authors in this book do not only want the oppressed groups they write about to be accepted and allowed to participate in what exists. They want to change existing environmentalism so that it recognises diverse approaches and achieves social and environmental justice.

An aspect that does not appear in the title of this book is equity, which refers to 'the absence of systematic disparities … between groups with different levels of underlying social advantage/disadvantage – that is, wealth, power, or prestige' (Chin and Chien, 2006, p. 79). Equity takes diversity and inclusion one step further. It can be differentiated from inclusion in that it considers outcomes at the systemic level and calls for the righting of systemic and structural injustices. Although the book focuses on diversity and inclusion as the initial steps, equity needs to be the ultimate aim. This is discussed in many of the chapters and, in addition, the final chapter of the book particularly considers how to move from diversity and inclusion to equity.

What is 'environmentalism'?

Debates continue about what the concepts of 'environment' and 'environmentalism' actually mean. It is clear that 'environment' can have a range of meanings and that 'environmentalism' comprises a wide range of perspectives and activities (Wardle et al., 2019; Newell, 2020). Environmental organisations and movements come in many forms and include a vast array of objectives and values, from eco-socialism and ecofeminism to nature conservation and green capitalism. However, for several decades, there have been analyses which see environmentalism, at least in the UK and US, as tending to be defined in terms that resonate with the dominant groups in society, such as White, male, middle-class groups (e.g. Agyeman, 2002; Allen et al.,

2007; Bell, 2020). For example, in the US, Allen et al. (2007, p. 124) noted that 'what the environment is (fragile ecosystem to be protected versus a place of dangerous threat)' differs dramatically between social groups. He contrasted the middle-class environmentalist, 'engaged in bird watching, recycling, "buying green", hiking', with the working-class environmentalist concerned with 'children playing in the shadow of smokestacks' (Allen et al., 2007, p. 124). The mainstream environmental movement has also tended to segregate 'environmental' from 'social' issues, as Di Chiro (2008, p. 279) highlights:

> Defining what counts as an environmental problem and what doesn't invites certain alliances and inhibits others, and the environmental movement has shot itself in the foot by adopting the definitional frontiers that delegate different issues as either inside or outside the environmental "frame". The conceptual-ideological mechanisms of exclusion and inclusion, which draw clear distinctions between problems that are defined as "social" (jobs, housing, transportation, public health, racial and sexual inequality, violence, poverty, reproductive freedom) and those termed "environmental" (global warming, natural resource conservation, pollution, species extinction, overpopulation) have led to the endless fragmentation of progressive movements.

Without being aware of how we are defining 'environmental issues', 'environmentalism', and what it is to be 'an environmentalist', we can be inadvertently discriminatory and disempowering of marginalised groups (Haluza-Delay et al., 2009; Schlosberg, 2007). To promote a just environmentalism, it will be important to include different ways of knowing about the environment and to incorporate diverse definitions (Schlosberg, 2004, 2007; Gosine and Teelucksingh, 2008; Haluza-Delay et al., 2009).

The consequence of excluding definitions is that when disadvantaged groups approach mainstream environmental organisations for support with their issues - for example, the location of hazardous facilities in their communities - their issues have in the past been deemed to be outside the scope of what is considered to be 'environmental'. For example, Di Chiro (1996) described how an environmental organisation refused to support a low-income community in their campaign to address problems caused by a local incinerator because 'the poisoning of an urban community by an incineration facility was a "community health issue", not an environmental one' (Di Chiro, 1996, p. 299). Hence, definitions invite inclusion and exclusion and facilitate certain alliances while undermining others. Environmentalists from diverse communities have, for some time, been bringing new vocabulary to environmentalism, such as 'ecological debt' and 'climate justice', which helps in the understanding of complex environmental problems and potential solutions.

In this book, we use the widest possible understanding of the term 'environmentalism' incorporating the notions of 'the environmentalism of everyday life'

(Pulido, 1998, p. 30) as well as 'wilderness and biodiversity'. We also, where relevant, continue to discuss what environmentalism is and could be.

Overview of the chapters

This book has been organised in terms of a range of equalities groups. However, in order to take into account the additional or different impacts of multiple oppressions, we speak of 'intersectionality' which emphasises that one identity cannot be experienced outside of other identities (Crenshaw, 1989). These intersectional issues are discussed, where relevant, in each chapter.

In Chapter 2, *Harriet Larrington-Spencer, Deborah Fenney, Lucie Middlemiss* and *Aleksandra Kosanic* discuss 'disabled environmentalisms'. They argue that the environmental movement has an ableist ethos and that there is a lack of recognition of the diverse needs and capabilities of disabled people. However, they go on to discuss the possibility for 'disabled environmentalisms': creating spaces for disabled engagement with this movement.

Chapter 3 by *Clara Greed* looks at women and environmental planning. She discusses how the women and planning movement has long argued that women's lives, travel patterns, responsibilities and experiences are different from those of men. The chapter critiques the extent to which the modern environmental planning movement is taking on board women's issues and proposes more inclusive alternatives.

In Chapter 4, *Silpa Satheesh* emphasises that environmental movements in the Global South have a very long history dating back to pre-colonial periods. Yet somehow this struggle is missing from the mainstream literature on environmentalism. The chapter underscores the need to decolonise the research on environmental movements by including poor and working-class environmentalism from the Global South.

Returning to the issue of social class, in Chapter 5, I discuss the exclusion of working-class people from the environmental movements of the Global North. This chapter outlines the barriers to engagement, giving examples in relation to some groups that are currently very active, such as Extinction Rebellion. It will also describe 'working-class environmentalism' as an important strand of environmentalism that builds on trade unionism and the environmental justice movement.

In Chapter 6, *Emma Foster* considers environmentalism in relation to the politics and activism related to lesbian, gay, bisexual, transgender, queer (or questioning), intersex, and asexual (or allies) and other relevant identities (LGBTQIA+). She highlights that while progressive politics is often conflated with commitments to LGBTQIA+ issues and ethical consideration of the natural environment, there are a number of tensions and contradictions. Ultimately though, the chapter demonstrates the radical potential of an LGBTQIA+ inclusive ecological/environmental politics.

Considering how environmental groups can better learn from communities of colour, in Chapter 7, *Roger Griffith* and *Gnisha Bevan*, draw on their experience of working with Bristol's Black and Green project. The chapter argues that obscuring the positive contributions of Black, Asian and Minority Ethnic communities is counterproductive: contributing to structural societal racism, global inequality and weakened environmentalism. It describes how the Black and Green project aimed to overcome this lack of recognition, working with communities of colour in low-income areas to create new leaders and challenge perceptions.

In Chapter 8, *Gary Haq* examines how environmental campaigns, policies and strategies have included and excluded older people. It reviews the evidence base, from early nature conservation to climate action in the UK, North America and globally, to consider how older people have been perceived, considered and included. The chapter makes recommendations for policy and practice and how to harness the power of this growing demographic group.

Chapter 9 by *Benjamin Bowman,* myself and *Becky Alexis-Martin* on young people looks at the exciting vibrancy of recent youth actions for the climate. The climate strikes brought global attention to the issues of climate change and reminded the world of the importance of incorporating the voices of youth into deciding on environmental policies and solutions. The chapter also discusses how young people are disempowered by society such that this tremendous energy and moral critique is held back.

In Chapter 10 on 'policies and change', I will draw together the issues highlighted in other chapters – looking at cross-cutting themes. The chapter will also focus on the policies that will drive forward an inclusive environmentalism, from the organisational to the societal level.

It is hoped then that this book will be both instructive and inspirational. Though some of it may make for difficult reading, where we see that we may have made mistakes in the past, it is important to face these in order to progress. The voices found in these pages are intended to unsettle and decolonise dominant narratives about environmentalism and marginalised people. We may not have all the answers but it is hoped that acknowledging, honouring and respecting these perspectives can lead to mutually reinforcing achievements for environmental and equalities communities.

Note

1 In this chapter, I have capitalised 'White' to signify that race is socially constructed and to de-centre and de-normalise 'white'. However, some authors in this book have used the non-capitalised term 'white' so as to distance from white supremacist groups that also capitalise it. For current debates on this, see Apiah (2020) and Brenner (2020).

References

Agyeman, J. (2002) 'Constructing environmental (in)justice: Transatlantic tales', *Environmental Politics*, vol 11, no 3, pp. 31–53.

Allen, K., Daro, V., and Holland, D.C. (2007) 'Becoming an environmental justice activist', in R. Sandler and P.C. Pezzullo (eds.), *Environmental Justice and Environmentalism: The Social Justice Challenge to the Environmental Movement.* MIT, Cambridge, MA.

Apiah, K.A. (2020) 'The case for capitalizing the B in Black', *The Atlantic*, 18 June.

Bell, K. (2020) *Working Class Environmentalism: An Agenda for a Just and Fair Transition to Sustainability*. Palgrave, London.

Bell, K. (2016) 'Bread and roses: A gender perspective on environmental justice and public health', *International Journal of Environmental Research and Public Health*, 10, 1005.

Brenner, E. (2020) *Black and White: When Should We Capitalize?* 3 August www.righttouchediting.com/2020/08/03/black-and-white-when-should-we-capitalize/.

Bryant, B. and Mohai, P. (eds.) (1992) *Race and Incidence of Environmental Hazards*. Westview Press, Boulder, CO.

Bullard, R. (1990) *Dumping in Dixie: Race, Class and Environmental Quality*. Westview Press, Boulder, CO.

Carmeli, A., Reiter-Palmon, R. and Ziv, E. (2010) 'Inclusive leadership and employee involvement in creative tasks in the workplace: The mediating role of psychological safety', *Creativity Research Journal*, vol 22, no 3, pp. 250–260.

Chin, M.H., and Chien, A.T. (2006) 'Reducing racial and ethnic disparities in health care: An integral part of quality improvement scholarship', *Quality and Safety in Health Care*, vol 15, pp. 79–80.

Cox, T. (1993) *Cultural Diversity in Organizations: Theory, Research and Practice*. Berrett-Koehler, San Francisco, CA.

Crenshaw, K. (1989) 'Demarginalizing the intersection of race and sex: A black feminist critique of antidiscrimination doctrine, feminist theory and antiracist politics', *University of Chicago Legal Forum*, vol 1989, no 1, pp. 139–167.

Cutter, S.L. and Solecki, D. (1996) 'Setting environmental justice in space and place: Acute and chronic airborne toxic releases in the Southeastern United States', *Urban Geography*, vol 17, pp. 380–399.

Di Chiro, G. (1996) 'Nature as community: The convergence of environment and social justice', in W. Cronon (ed.), *Uncommon Ground: Rethinking the Human Place in Nature*. W.W. Norton, New York, NY.

Di Chiro, G. (2008) 'Living environmentalisms: Coalition politics, social reproduction, and environmental justice', *Environmental Politics*, vol 17, no 2, pp. 276–298.

Downey, S.N., van der Werff, L., Thomas, K.M., and Plaut, V.C. (2014) 'The role of diversity practices and inclusion in promoting trust and employee engagement', *Journal of Applied Social Psychology*, vol 45, no 1, pp. 35–44.

Echols, A. (1989) *Daring to be Bad: Radical Feminism in America, 1967–1975*. University of Minnesota Press, Minneapolis, MN.

Feliz Brueck, J. and McNeill, Z. (eds.) (2020) *Queer and Trans Voices: Achieving Liberation through Consistent Anti-Oppression*. Sanctuary Publishers, Kindle.

Ferdman, B.M. (2014), 'The practice of inclusion in diverse organizations: toward a systemic and inclusive framework', in Ferdman, B.M. and Deane, B.R. (Eds), *Diversity at Work: The Practice of Inclusion*, Wiley-Blackwell, Oxford, pp. 3-54.

Ferdman, B.M. (2017) 'Paradoxes of inclusion: Understanding and managing the tensions of diversity and multiculturalism', *The Journal of Applied Behavioral Science*, vol 53, no 2, pp. 235–263.

Finney, C. (2004) 'Can't see the black folks for the trees', in S. Weir and C. Faulkner (eds.), *Voices of a New Generation: A Feminist Anthology*. Pearson, New York.

FoE (2001) *Pollution and Poverty – Breaking the Link*. Friends of the Earth, London.

Fraser, N. (1995a) 'Reframing justice in a globalizing world', *New Left Review*, vol 36, pp. 1–19.

Fraser, N. (1995b) 'Politics, culture, and the public sphere: Toward a postmodern conception', in L. Nich and S. Seidman (eds.), *Social Postmodernism: Beyond Identity Politics*. Cambridge University Press, Cambridge.

Fraser, N. (1992) 'Rethinking the public sphere: A contribution to the critique of actually existing democracy', in C. Calhoun (ed.), *Habermas and the Public Sphere.* MIT Press, Cambridge, CA.

Gitlin, T. (1995) *The Twilight of Common Dreams: Why America Is Wracked by Culture Wars.* Metropolitan Books, New York, NY.

Gosine, A. and Teelucksingh, C. (2008) *Environmental Justice and Racism in Canada: An Introduction.* Emond Montgomery Publications, Toronto.

Habermas, J. (1987) *The Theory of Communicative Action: Volume Two.* Beacon Press, Boston, MA.

Haluza-Delay, R., O'Riley, P., Cole, P., and Agyeman, J. (2009) 'Speaking for ourselves, speaking together: Environmental justice in Canada', in J. Agyeman, P. Cole, R. Haluza-Delay and P. O'Riley (eds.), *Speaking for Ourselves: Environmental Justice in Canada.* UBC Press, Vancouver.

Harper, B. (2009) *'Why don't black people go camping…?' Critical whiteness studies in environmental education.* http://sistahvegan.com/2009/07/17/why-dont-black-people-go-camping-critical-whiteness-studies-in-environmental-education/.

Heiman, M. (1996) 'Race, waste, and class: New perspectives on environmental justice', *Antipode,* vol 28, pp. 111–121.

Hirak, R., Peng, A.C. Carmeli, A. and Schaubroeck, J.AM. (2012) 'Linking leader inclusiveness to work unit performance: The importance of psychological safety and learning from failures', *The Leadership Quarterly,* vol 23, no 1, pp. 107–117.

Hossain, M., Atif, M., Ahmed, A., and Mia, L. (2019) 'Do LGBT workplace diversity policies create value for firms?', *Journal of Business Ethics,* vol 97, no 4, pp. 1–17.

Hunter, J.D. (1991) *Culture Wars: The Struggle to Define America.* Basic Books, New York, NY.

IPBES (2019) *Summary for Policymakers of the IPBES Global Assessment Report on Biodiversity and Ecosystem Services.* https://ipbes.net/sites/default/files/inline/files/ipbes_global_assessment_report_summary_for_policymakers.pdf.

Intergovernmental Panel on Climate Change (2018) *Global Warming of 1.5 °C.* www.ipcc.ch/report/sr15/.

Jackson, S.E., and Joshi, A. (2011) 'Work team diversity', in S. Zedeck (ed.), *APA Handbook of Industrial and Organizational Psychology.* American Psychological Association, Washington DC.

Labonte, R. (2004) 'Social inclusion/exclusion: Dancing the dialectic', *Health Promotion International,* vol 19, no 1, pp. 115–121.

Mannix, E., and Neale, M.A. (2005) 'What differences make a difference? The promise and reality of diverse teams in organizations', *Psychological Science in the Public Interest,* vol 6, no 2, pp. 31–55.

Mansbridge, J. (1999) 'Should blacks represent blacks and women represent women? A contingent "yes"', *Journal of Politics,* vol 61, no 3, pp. 628–657.

Milojevic, A., Niedzwiedz, C.L., Pearce, J., Milner, J., MacKenzie, I.A., Doherty, R.M., and Wilkinson, P. (2017) 'Socioeconomic and urban-rural differentials in exposure to air pollution and mortality burden in England' *Environmental Health,* vol 16, no 104, pp. 1–10.

Mor Barak, M.E., Lizano, E.L., Kim, A., Duan, L. Rhee, M. Hsiao, H., and Brimhall, K. (2016) 'The promise of diversity management for climate of inclusion: A state-of-the-art review and meta-analysis', *Human Service Organizations,* vol 40, no 4, pp. 305–333.

Mor-Barak, M.E., and Cherin, D.A. (1998) 'A tool to expand organizational understanding of workforce diversity: Exploring a measure of inclusion-exclusion', *Administration in Social Work,* vol 22, no 1, pp. 47–64.

Morley, R. (2006) 'The cost of being poor: Poverty, lead poisoning, and policy implementation', *The Journal of the American Medical Association,* vol 295, no 14, pp. 1711–1712.

Morello-Frosch, R., Pastor, M., and Sadd, J. (2002) 'Environmental justice and southern California's "riskscape": The distribution of air toxics exposures and health risks among diverse communities', *Urban Affairs Review*, vol 36, no 4, pp. 551–578.

Newell, P. (2020) *Global Green Politics*. Cambridge University Press, Cambridge.

Nishii, L. H. (2013) The benefits of climate for inclusion for gender-diverse groups. *Academy of Management Journal*, vol 56, no 6, pp. 1754–1774.

Palamar, C. (2008) 'The justice of ecological restoration: Environmental history, health, ecology, and justice in the United States', *Human Ecology Forum*, vol 15, no 1, pp. 82–94.

Pellow, D., and Park, L. (2003) *The Silicon Valley of Dreams: Environmental Injustice, Immigrant Workers, and the High-Tech Global Economy*. New York University Press, New York, NY.

Prakash, S. (2007) 'Beyond dirty diesels: Clean and just transportation in northern Manhattan', in R. Bullard (ed.), *Growing Smarter: Achieving Livable Communities, Environmental Justice, and Regional Equity*. MIT Press, Cambridge, MA.

Pulido, L. (1998) 'Development of the 'People of Color' identity in the environmental justice movement of the Southwestern United States', *Socialist Review*, vol 26, no (3–4), pp. 145–180.

Schlosberg, D. (2004) 'Reconceiving environmental justice: Global movements and political theories', *Environmental Politics*, vol 13, no 3, pp. 517–540.

Schlosberg, D. (2007) *Defining Environmental Justice: Theories, Movement and Nature*. Oxford University Press, Oxford.

Shore, L.M., Cleveland, J.N. and Sanchezc, D. (2018) 'Inclusive workplaces: A review and model', *Human Resource Management Review*, vol 28, no 2, pp. 176–189.

Shore, L.M., Randel, A.E., Chung, B.G., Dean, M.A., Ehrhart, K.H. and Singh, G. (2011) 'Inclusion and diversity in work groups: A review and model for future research', *Journal of Management*, vol 37, pp. 1262–1289.

Snow, D. and McAdam, D. (2000) 'Identity work processes in the context of social movements', in S. Stryker, T.J. Owens and R.W. White (eds), *Self Identity and Social Movements*. University of Minnesota Press, Minneapolis, MN.

Steffen, W., Richardson, K., Rockström, J., Cornell, S.E., Fetzer, I., Bennett, E.M. et al. (2015) 'Planetary boundaries: Guiding human development on a changing planet', *Science*, vol 347, no 6223, 1259855.

Sze, J. and London, J. (2008) 'Environmental justice at the crossroads', *Sociology Compass*, vol 2, no 4, pp 1331–1354

Tang, Y., Jiang, C., Chen, Z., Zhou, C., Chen, C. and Yu, Z. (2015) 'Inclusion and inclusion management in the Chinese context: An exploratory study', *The International Journal of Human Resource Management*, vol 26, no 6, pp. 856–874.

Tarrow, S. (1998) *Power in Movement: Social Movements and Contentious Politics*. Cambridge University Press, Cambridge.

Taylor, D. (1993) 'Environmentalism and the politics of inclusion', in R. Bullard (ed.), *Confronting Environmental Racism: Voices from the Grassroots*. South End Press, Boston, MA.

Taylor, D. (2000) 'The rise of the environmental justice paradigm: Injustice framing and the social construction of environmental discourses', *American Behavioral Scientist*, vol 43, no 4, pp. 508–580.

United Church of Christ (2007) *Toxic waste and race at twenty: 1987–2007*. www.ucc.org/.

United Church of Christ Commission for Racial Justice (1987) *Toxic wastes and race in the United States*. www.ucc.org/.

Walcott, R. (2019) 'The end of diversity', *Public Culture*, vol 31, no 2, pp. 393–408.

Walker, G., Fairburn, J., Smith, G. and Mitchell, G. (2003) *Deprived Communities Experience Disproportionate Levels of Environmental Threat*. Environment Agency, Bristol.

Wardle, P., Robin, L., and Sverker S. (2019) *The Environment: A History of the Idea*. Johns Hopkins University Press, Baltimore, MD.

Weldon, S.L. (2006) 'Inclusion, solidarity, and social movements: The global movement against gender violence', *Perspectives on Politics*, vol 4, no 1 pp. 55–74.

Williams, M. (1998) *Voice, Trust and Memory: Marginalized Groups and the Failings of Liberal Representation*. Princeton University Press, Princeton, NJ.

WWF (2020) 'The Living Planet Report 2020', World Wide Fund for Nature, Woking.

Young, I.M. (2000) *Inclusion and Democracy*. Oxford University Press, Oxford.

2

DISABLED ENVIRONMENTALISMS

Harriet Larrington-Spencer, Deborah Fenney,
Lucie Middlemiss and Aleksandra Kosanic

Introduction

Life has changed unrecognisably since the Industrial Revolution and, since the 1950s, rapid economic growth has brought us increased consumption and globalisation. Taking into account these multiple challenges, and in order to leave a resilient planet for generations to come, we need to focus on questions of equality/equity and sustainability. This will enable us to achieve the Agenda 2030 Sustainable Development Goals (SDGs) and to leave no one behind (Leach et al., 2018; UN, 2018). As Leach et al. (2018, p. 1) stated, '[i]t is no longer possible nor desirable to address the dual challenges of equity and sustainability separately'.

Within policy and practice addressing environmental problems there is increasing recognition that inclusive action, accounting for people's different needs and capabilities, is necessary when planning for change (Middlemiss, 2018). We can see evidence of this recognition in the actions of international monitoring and decision-making bodies, such as the Intergovernmental Panel on Climate Change (IPCC) and the Intergovernmental Science-Policy Platform on Biodiversity and Ecosystem Services (IPBES), which are charged with providing comprehensive evidence of anthropogenic changes. These institutions now consider the impacts of such environmental changes on diverse populations, including indigenous people, women and older people (IPCC 2018; IPBES 2019). However, the impacts of environmental change on disabled populations are not yet fully recognised (Kosanic et al., 2019a). Within academic literature there is increasing attention to the significance of some forms of inequality in considering how to address environmental issues. Particularly notable examples have been the importance of income (Büchs and Schnepf, 2013), class (Johnston, 2008; Bell, 2020) and gender (Hawkins, 2012; MacGregor, 2016) in shaping people's experiences of environmentalism. This book, indeed, is a welcome effort in bringing these contributions into conversation.

However, while the environmental justice movement refers to 'all people's needs', it rarely addresses disabled people's needs specifically (Johnson, 2011) and disability rarely appears in either the environmental agenda or as a category of analysis in academic research on environment (Kafer, 2013). As such, the intersectional issues that disabled people face (through gender, race, class sexuality, etc.) are largely ignored in theory and action. This is particularly surprising considering that the United Nations report on disability and the SDGs released in 2018 evidenced interlinkages between disability and several SDGs and highlighted inequalities (Wilkinson and Pickett, 2009; Brondizio et al., 2016; UN, 2018). It is also notable that the emerging, albeit limited, literature that includes disabled populations shows that disabled people experience different impacts from changing environments (Morris et al., 2018; Lunga et al., 2019; Watts et al., 2018; Kosanic and Petzold, 2020). It also shows that they face different barriers to inclusion and have different opportunities to contribute to the agenda (Hemingway and Priestley, 2006; Leipoldt, 2006; Charles and Thomas, 2007; Imrie and Thomas, 2008; Wolbring, 2009; Alaimo, 2010; Fenney and Snell, 2011; Withers, 2012; Fenney 2017; Fenney Salkeld, 2016, 2019).

In this chapter, we develop an agenda for the fostering of disabled environmentalism, demonstrating how such an agenda, through the centralising of disabled bodies, can enable environmental justice movements to be truly inclusive of all people's needs. In doing so, we recognise that understanding disabled people's experiences as citizens, activists and subjects of environmental policies and interventions is a necessary first step in understanding the development of such an agenda. We further acknowledge that we need an interdisciplinary research agenda, bringing together geography, ecology and environmental social sciences with disability studies. To achieve this, we begin by identifying the barriers that disabled people face to inclusion in environmentalism, both in terms of integrating environmentalism in everyday practices and in participating in the organisation and implementation of more public forms of environmentalism, such as organised activism and protest. These barriers are then contextualised within academic scholarship on disability and environmentalism and through the social and medical models of disability. In the final section, we propose an agenda of 'cripping' environmentalism, centring disability within environmental futures, recognising the disabled body as the antithesis of the capitalist modes of production upon which environmental degradation is predicated, and demonstrating that an accessible world is not only desirable but also necessary for environmental protection and social justice.

Barriers to disabled people's inclusion in environmentalism

In this section, we provide a broad overview of the barriers that disabled people can experience when attempting to participate in environmentalism – both in everyday environmentalism as well as in the organisation and implementation of more public forms of activism. Our list is by no means exhaustive, but we select these to demonstrate the wide range of barriers and in order to highlight how endemic they are within environmentalism. By doing so, this section provides both the context for demonstrating the necessity of developing disabled environmentalisms, as well as a space for readers to critically reflect on the inclusivity of their own environmental practices.

Before we begin, we should note that disabled people's experiences are hugely diverse and disabled people have varying politics, expectations and desires to be engaged in an environmental agenda. The concept of disability is very broad, and can relate to a wide range of impairments, including autism spectrum conditions, long-term health conditions, mental health conditions, physical or mobility impairments, sensory impairments (e.g. deafness and blindness) and learning difficulties. Here, when we identify barriers, we note that these will not be barriers for all disabled people. However, we include these because we believe that they will be substantial barriers for some. We also recognise that many of the barriers highlighted in this section are specific to the United Kingdom (UK). This is because they represent many of the authors' lived, and lived alongside, experiences of disability. Their context specificity, in itself, demonstrates the need for a deeper and wider engagement with disability within the environmental agenda.

Material barriers

The body

People can be excluded from environmentalist practices due to their material circumstances, sometimes including physical exclusion for bodily reasons. Some bodies can easily travel by bike, others cannot; some can exist off-grid, some cannot; some can reduce the water and energy consumption associated with cleanliness practices, some cannot (Fenney and Snell, 2011; Fenney, 2017; Fenney Salkeld, 2016, 2019). Environmental campaigns often target what are considered as 'convenience items' – for example, plastic straws, prepackaged produce and wet wipes. However, what these campaigns often fail to consider is that items considered as 'convenient' by non-disabled people can often be central to the health and independence of disabled people. Bendy plastic straws can be essential for drinking, and drinking at the correct angle, for some mobility-impaired people and single-use straws are often much more hygienic than reusable straws. Pre-chopped produce supports people with limited hand dexterity as well as energy-related issues to maintain independence in cooking. Wet wipes can be critical to support some disabled people to have control over their own hygiene.

This is also not simply about the physicality of bodies. Mental health can also be a material factor in exclusion. Mental health problems can lead to fatigue which necessitates using more energy-intensive modes of transport. People can face difficulties using crowded public transport due to panic attacks, making car use a necessity. Furthermore, medication, for either mental or physical health, is sometimes stigmatised by environmentalists as coming from 'big pharma' (Fenney Salkeld, 2019).

The built environment

Material circumstances and barriers to environmentalism relate not only to the materiality of the body but also to the materiality of living. Barriers to environmentalism can be in terms of meetings being held in rooms only accessible by stairs (Fenney, 2017). This is common as much environmental activism is unfunded and free meeting spaces are less likely to have access facilities, such as ramps and lifts. Spaces of environmental activism, such as public protests, frequently do not have Changing Places (fully accessible toilet facilities), or even any toilet facilities, within close proximity.

Social barriers

Given that disability is rarely discussed in relation to environmentalism, it is not surprising that campaigns to promote active travel or banning plastic straws do not take into account the needs of disabled people. Worse than this is the creation of stigma towards disabled people (for not acting 'correctly') through environmental campaigning. Campaigns that call for a total ban on items that enable disabled people to live their lives (e.g. plastic straws, cars, etc.) not only have health and independence implications for disabled people, but can also label disabled people as anti-environmental, and result in public shaming and policing of their actions. For instance, the widespread campaign against single-use plastic straws in 2019 has resulted in these no longer being readily available in public places. This puts those disabled people who need to use straws to drink at risk, as well as making it more risky to be disabled and drinking in public.

Note that stigmas associated with disability are widespread in society, with many people holding disableist (prejudice against disabled people) or ableist (an assumption of non-disability) views (Chapman, 2020; Renke, 2020). Such prejudiced viewpoints can also be traced through the environmentalist movement, which tends to place a high social value on self-sufficiency, the use of 'natural' products (as opposed to the medication and technology that keep some disabled people alive), and the strength of the human body in nature, for instance (Fenney, 2017; Ray and Sibara, 2017). The expectation by the movement that environmental activists will be non-disabled is, in itself, likely to be stigmatising, and to create additional barriers to participation.

Institutional barriers

Participation in environmentalism, either through everyday practices or within activist organisations, movements and demonstrations, is often perceived as indicative of non-disability by punitive institutions, such as the police, or state departments, for instance, the Department for Work and Pensions (DWP). Disabled people's participation in environmentalism can be considered evidence of exaggeration or fictionality of disability and thus require regulation and disciplinary action. For example, in 2018, Lancashire Police admitted passing on details and video footage of disabled anti-fracking protesters to the DWP, considering ambulatory wheelchair use a 'clear suggestion that fraud may be being committed' (Rahim, 2018). This resulted in a number of disabled anti-fracking activists being contacted by the DWP for benefits reassessments (Pring, 2018). More recently, it has emerged that Greater Manchester Police has a written agreement with the DWP to pass on the details of any disabled person taking part in protests in the region (Pring, 2019).

Breaches of disabled people's right to protest create a culture of fear, where legitimate engagement in environmental activism risks illegitimate sanctions (Pring, 2018, 2019). Similarly, the link between physical activity and assumed physical ability can limit disabled people's participation in active transport. In their annual survey of disabled cyclists, Wheels for Wellbeing (2018) found that 49% of respondents were worried that their benefits would be withdrawn or reduced as a result of being physically active, and 17% reported that this worry deterred them from cycling or caused them to cycle less or give up cycling. These concerns are legitimate – 6% reported that they have had benefits reduced or withdrawn as a result of cycling and being physically active.

Financial barriers

Being disabled is expensive. Disabled people face extra costs in order to have the same standard of living as non-disabled people (Scope, 2019). These include higher transport costs, expensive equipment, paying more for housing and paying more for energy. For disabled adults these costs amount to an average of half the household income (after housing costs). Disabled adults are also more likely to be unemployed and more likely to live in poverty than non-disabled adults (Scope, 2019), exacerbating financial pressure for many disabled people. Disability and income poverty are strongly correlated and, as such, households can be affected by a double constraint. This is an issue of distributive justice and also impacts on households' access to participation in some environmentally friendly practices. Equally, income is closely related to environmental impact (Oswald et al., 2020), with people on higher incomes having higher impacts, and as such disabled people are likely to have smaller impacts on average.

A classic example of reducing environmental impact in everyday life is replacing private car use with emission-free travel such as cycling. Cycling as active travel for disabled people is a particularly promising lower-impact transport solution considering that many disabled people find cycling easier than walking (Wheels for Wellbeing, 2019). However, while standard two-wheel bicycles can be purchased relatively cheaply and are widely available second-hand, the same cannot be said for non-standard and adapted cycles which disabled people may need. These include equipment such as trikes, handcycles, recumbents and electric cycles, the cost of which can easily extend into thousands of pounds. In addition to the higher costs of purchase, there are also cost implications in terms of secure storage and insurance due to their larger size and higher value. In the UK, while the mobility component of disability benefits can be used for Motability and private car access, there is no provision for covering non-standard cycles. This means that disabled people have to cover the higher cost of accessing more environmentally friendly travel themselves.

Temporal barriers

Another example of reducing environmental impact in everyday life is replacing private car use with public transport, such as buses and trains. However, there are implications in terms of time added as a result of inaccessibility. In the UK, while the majority of buses are now accessible for people with mobility aids, there is often only one wheelchair space per bus. Waiting time is then extended for a person who requires this space if it is occupied by another disabled person or by non-disabled users and is not being regulated by the driver. As the majority of trains and stations in the UK are not designed for level boarding, ramp access is needed for mobility-impaired people. However, in addition to ramp access being steep and sometimes dangerous, mobility-impaired disabled people are required to pre-book and depend upon station assistance. This not only reduces flexibility in travel but also has significant implications when station assistance fails to materialise and disabled people are unable to (dis)embark. These failures in support provision increase journey times as well as create significant frustration, making public transport an inconvenient form of travel compared to private vehicles.

Public transport use during the coronavirus (COVID-19) pandemic is also riskier for people who are particularly vulnerable to the effects of the virus. External shocks such as these have particularly pronounced effects on disabled people. This is why it is doubly important to pay attention to the potentially regressive effects of environmental policy and practice on disabled people – there can be less flexibility in the way that people are able to cope with changes in their lives. This is compounded given that disabled people earn less (as above), and these inequalities will be even more exacerbated in the Global South. Hence, we can also see how mobility can be (further) restricted if policy increases the cost of car use without protecting or excluding people with mobility impairments.

The personal costs – being the disabled killjoy

Being involved in environmental activism, just as being involved in any social activity, is likely to bring disabled people into contact with ableist and disableist views (as in 'social barriers' above). Given the limited recognition of disabled people in the environmental movement, there is extra labour in being disabled as an environmental activist. Symbolic exclusion and violence have an important impact on activists, which are sometimes difficult to document. How do we measure non-participation or offence caused by particular framings of environmentalism? Disabled people who wish to participate often find that the impetus is upon themselves to highlight and resolve access issues. The labour of this is time-consuming, preventing disabled people from participating more fully in environmental activism itself. To undertake such labour is also exhausting and one can both be worn down and become down when having to continuously push against what has already been constructed with little regard for diversity (Ahmed, 2017). And through this process, the disabled body – building on the work of Ahmed (2017) – becomes the 'killjoy'. By highlighting inequalities in accessibility, as well as the specific challenges associated with a disabled-inclusive environmentalism, disabled people kill joy (Brooks and Snelling, 2018). By highlighting inequalities in accessibility, disabled people may be considered 'wall makers' (Ahmed, 2017, p. 252) – drawing attention away from the environmental concerns central to the activist movements themselves. Disabled people who call out the symbolic exclusions they witness may be met with denial or accused of not caring about the environment or of damaging or distracting from movement aims, thus further alienating them from environmental activism.

Disability and environmentalism in the academic context

As previously mentioned, to date there has been limited work that directly addresses environmentalism and disability. In this section we outline the 'so-far' contributions to this field, recognising opportunities to develop this work through links to the broader critical body of work on social inequalities and justice in environmentalism documented in this book. We look below at how disability studies understand the environment and how environmental studies understand disability and then outline more recent engagements, recognising the necessity for more critical and transformative approaches.

Disability studies on the environment

While the environment is a central concern within disability studies, the focus has almost exclusively been on the built environment and ensuring that the built environment is accessible to disabled people (Kafer, 2013; Ray and Sibara, 2017) – in other words, overcoming 'material barriers', some of which we have identified above, and ensuring an enabling built environment for the materiality of disabled bodies. A few examples of what overcoming material barriers involves would be – providing level-access or ramp access to buildings, disabled parking spaces, Changing Places toilet facilities, dropped kerbs and signal-controlled pedestrian crossings.

This focus on the built environment within disability studies is situated within the social model of disability. The social model of disability recognises that many of the barriers to full and meaningful participation in society that disabled people face would not exist if the physical and social environments in which people live were adapted to accommodate a diverse range of embodiments and interactions (Thomas, 1999). Such barriers within the built environment, as well as social barriers, institutional barriers and financial barriers, are all discussed above within the context of environmentalism. The social model recognises that, as a result of barriers external to disabled people themselves, disabled people face a form of social oppression, that is, disablism. A related concept useful to this work is 'ableism'. Ableism describes how society is generally organised from a non-disabled viewpoint, and thus ignores or minimises other experiences and embodiments. An unquestioning able-bodied 'normal' implicitly positions disabled people as 'abnormal' (Imrie, 1996), rather than addressing the assumptions that underlie how society has constructed what normal is deemed to be. Ableism underlies many of the issues disabled people face when trying to engage with environmental movements (Fenney, 2017), in addition to any overt disablism they may face.

Environmental studies and disability

Within environmental scholarship there is engagement with disability, most commonly in terms of public health and the impact of environmental degradation on disabled bodies or disabled bodies as the outcome of degraded environments (Ray and Sibara, 2017). For example, climate change is expected to disproportionately affect disabled people (Smith et al., 2017; Kosanic et al., 2019a). The higher intensity and frequency of extreme weather events, such as cyclones, hurricanes, extreme precipitation events, heatwaves, cold waves, droughts and storm surges, will affect disabled people more severely (Beniston and Stephenson, 2004; Zwiers et al., 2013; Walsh-Warder, 2016; Baker et al., 2017). Slow onset impacts such as sea level rise, or impacts on ecosystems, such as changes in the geographical distributions of species including latitudinal, longitudinal and altitudinal changes, or changes in phenology, genetic diversity, species declines, invasive species intrusions and extinctions also might pose a greater threat to disabled people (Parmesan and Yohe, 2003; Kosanic et al., 2019b). In the epoch of the Anthropocene, climate and environmental change can exacerbate underlying conditions (e.g. through the mental and physical health impacts of extreme climatic events or pandemic situations) or can make disabled people susceptible to disaster-related injuries, displacement, vector- or water-borne diseases (e.g. malaria, EVD4, Zika) and potentially lead to higher mortality (Whitmee et al., 2015; Watts et al., 2018). Given that climate change poses long-term pressures to ecosystems and societies, impacting human well-being, migration, infrastructure and settlements, further impact on disabled people will be important to research, monitor and to take into account in policy (Bell et al., 2020).

For example, it is known that Small Island states have already experienced devastating climate change impacts due to sea level rise (Ourbak and Alexandre, 2018; Petzold and Alexandre, 2019). Furthermore, populations closely connected to nature and whose livelihoods are dependent on nature (i.e. indigenous and local populations) are at the highest risk in relation to climate change. Within indigenous and local communities, disabled populations, particularly children and women, will be the most vulnerable (Omolo and Mafongoya, 2019).

Without the sustainable use of nature, quality of life and human well-being is jeopardised and can further deteriorate the human benefits from nature (Martín-López et al., 2012; IPBES, 2019). These services have been described as 'ecosystem services' (ES) and defined as benefits that humans gain from the rest of nature – provisioning (e.g. food, timber and water), supporting (e.g. soil formation and retention, nutrient cycling and water cycling), regulating (e.g. local climate and air quality, pollination, carbon sequestration and storage) and cultural (e.g. spiritual, place attachment, recreational and aesthetic). These functions of services are essential for human quality of life and well-being (Liu et al., 2010; MEA, 2005; Kosanic and Petzold, 2020). A recent new concept of nature contributions to people (NCP) evolved aiming to include different values and stakeholders in order to better understand human–nature relationships (Diaz et al., 2018; IPBES, 2019). Similarly to ES, the NCP are organised into three broad groups: regulating (e.g. regulating soil or climate), material (e.g. food and energy) and non-material (e.g. recreation, inspiration and spiritual) (Díaz et al., 2019; IPBES, 2019, Pascual et al., 2017).

Provisioning and demand of NCP is, in many cases, context specific and could differ among marginalised groups (e.g. indigenous or disabled populations). For example, provisioning or demand for non-material NCP can be different for someone with a sensory impairment (i.e. relating to hearing or vision). Material NCP (food in particular) may not be evenly distributed among disabled and non-disabled children. This may result in malnutrition in disabled children (Adams et al., 2012; Kerac et al., 2014). In order to reach a just and sustainable future, we need to secure a broad range of NCP to meet needs and achieve a good quality of life (MEA, 2005; Diaz, 2020).

A second body of work in environmental studies centralises environmental exposure, toxicity and subsequent body burdens (Kafer, 2013; Ray and Sibara, 2017). Within this work, disability is often conflated with injustice (Kafer, 2013), with the disabled body, arising from such toxic environments, then commonly utilised to motivate public environmental movements (Di Chiro, 2010). There are parallels here with the framing of the wilderness as a place in which strong (male) bodies overcome adversity, and in which disability is seen as a threat, or risk to these strong bodies, arising from accidents in wild places (Ray, 2013). In both cases, the disabled body is positioned as a cautionary tale, an outcome of accident, or environmental degradation, rather than a whole human being with interests in environmental damage beyond their own body.

Positioning disabled bodies as unfortunate outcomes of human or natural environments echoes the medical model of disability, which is rooted in the discourse of disability as 'tragedy', and which offers only individualised solutions such as rehabilitation or cure to help a disabled person overcome their impairment (Oliver, 1990). While rehabilitation is not intrinsically negative, its misapplication can lead to people being forced into narrow 'able' embodiments and the devaluing of others. This model was the dominant way disability was viewed until the second half of the 20th century. In the 1960s and 1970s, groups of disabled activists began to develop an alternative model – known as the social model, set out above – to better describe their experiences.

The environmental movement, particularly in the English-speaking world, initially characterised environmental problems as stemming from overpopulation – too many people consuming too much stuff results in environmental damage. This argument has lost its dominance since the 1980s, as an alternative explanation came to the fore – too much stuff being consumed by the wealthiest results in environmental damage (Wingate, 2019). The 'overpopulation', or environmental Malthusianist, argument is still apparent in some environmentalist and conservationist responses to environmental problems, however. This perspective on environmentalism is important because of its link with the solution of 'population control', which, in itself, has strong links to eugenics thinking and indeed eugenics policy. It is an easy step from 'there are too many people' to 'there are too many of the wrong kind of people', and some of the foundational thinkers in this field had racist, classist and/or disableist views – read, for instance, Mildenberger's (2019) reflections on the influence of racism, eugenics and islamophobia on the work of Garrett Hardin. These views had very real consequences in policy: while we cannot say that environmentalism itself resulted in forced sterilisations of black, poor and disabled people, Ray (2013) documents how environmentalist ideas contributed to a world in which these things were possible. This is because the aversion to certain kinds of bodies (arising from environmental harm) draws problematic distinctions between those who belong to nature and those who are out of place. Furthermore, while explicitly racist, classist and disableist views in the movement are rarely acknowledged by its members, there are ongoing associations between the environmental movement and White, middle-class, non-disabled identities that can produce exclusion and prejudice. For instance, as we noted above, in describing social barriers to disabled people's engagement in environmentalism, the characterisation of environmentalism as embracing 'nature' and rejecting 'technology' can be problematic for disabled lives in which people rely on technology or medication for survival. The lack of consideration of disabled people is also likely to result in additional barriers to disabled people accessing, and exercising in, nature, both of which can have positive impacts on well-being. In addition, the conversations around 'needs' and 'wants', so widespread in the environmental literature, can suggest that only specific levels of 'needs' are acceptable, thereby excluding those who have specific physical needs in order to be able to exist in the world (Ahmed, 2010; Middlemiss, 2018).

Engaging with recent developments in disability studies

There is limited environmental scholarship that critically engages with disability studies itself (Ray and Sibara, 2017), and such discussions of integrating the social or medical models have been largely ignored in environmental thinking. New visions are built on the understanding that disability is the result of a complex interaction between individual, biological and social factors. One approach highlights that while people may have impairments that cause difficulties such as pain or fatigue, the disability they face is caused by unequal social relations with non-disabled people (Thomas, 1999). Another approach from the World Health Organization (WHO) suggests that disability should not be seen through either a 'medical' or 'social' model, uniquely, and advocates for a more balanced approach (WHO, 2011). These latter two approaches have advantages in recognising the nuances of disabled people's experiences, and also allow consideration of other factors that will influence a person's experience of disability, such as their gender, class and race. For example, someone who is able to pay for the most advanced adaptations and support will face fewer barriers, while someone who is disabled and Black is likely to face intersecting oppressions due to the impacts of both disablism and racism (Chahal, 2004).

Such approaches remain normative, not transformative. However, disabled people are excluded in many parts of society, and environmentalism is not necessarily exceptional in this regard. Exclusion is still endemic. Instead, what is needed is 'analyses that recognize and refuse the intertwined exploitation of bodies and environments without demonizing the illnesses and disabilities, and especially the ill and disabled bodies, that result from such exploitation' (Kafer, 2013, p. 158). Environmentalism and environmental scholarship could use some 'world making' (Goodman, 1978), understood as 'building and reshaping the realities in which we live' (Ginsberg and Rapp, 2017, p. 184). Such world-making processes would look at how disability worlds come into being and give glimpses of what can happen when disabled people are fully included in civic space (Ginsburg and Rapp, 2017), rather than considering disabled people as signifying 'the future of no future' (Kafer, 2013, p. 34). Such futures move beyond inclusion, as inclusion focuses upon integration into the normative, ableist, capitalist system without transformative change.

Cripping environmentalism

Centralising disability in environmental futures

Utilising a queer, feminist lens to build upon the social model of disability, recognising its limitations in terms of the marginalisation of disabled people who seek medical intervention and the reification of distinction between 'disabled' and 'non-disabled', Kafer (2013) suggests a political/relational model of disability. By doing so, Kafer seeks to make room for activism and transformative change, 'seeing disability as a potential site for collective reimagining' (Kafer, 2013, p. 9). Such a disability politics would be enabled not because of any essential similarities between disabled people, but because 'all have been labeled as disabled or sick and have faced discrimination as a result' (Kafer, 2013, p. 11). We use this political/relational model as a means through which to think about, and crip, environmentalism. Cripping environmentalism would involve centralising disability within environmental futures and demanding that an accessible world is possible and desirable (McRuer, 2006).

According to McRuer (2006), disability is the object against which imagined futures within capitalist, globalised and neo-liberal futures are produced. The very presence of the disabled body within these futures would signify a failure of these ideologies (McRuer, 2006), as it is rehabilitated, cured bodies that are considered 'the sign of progress, the proof of development' (Kafer 2013, p. 28).

These normatively curative imagined futures are founded in histories of ableism and disability oppression (Kafer, 2013) and commonly involve invocations of the able-bodied and able-minded White child as 'the fantasmic beneficiary of every political intervention' (Edelman, 2004, p. 2).

Processes of adverse environmental change and degradation, including but not limited to climate change, deforestation, air pollution, soil degradation, and land and water contamination, are widely attributed to capitalist systems of production (e.g. Magdoff and Foster, 2011; Bell, 2014; Klein, 2014). Capitalism is predicated on the perception of the natural environment as a space in which to expand and exploit, as capitalist systems of production require ever greater levels of consumption to not only drive the economy, but to keep it growing. Consequently there is increasing recognition that environmentalism and environmental justice movements need to be anti-capitalist.

However, environmentalism often involves imagined futures produced as oppositional to the disabled body (Kafer, 2013). The White, non-disabled child of capitalist futures is reproduced within environmental futures and aptly demonstrates how, within environmentalism, 'discourses of reproduction, generation and inheritance [which are] shot through with anxiety about disability' (Kafer, 2013, p. 29). This is particularly relevant considering that much of the focus on disability within environmental studies has been the disabling impacts upon bodies (Ray, 2013). As environmentalism is then theorised according to the abilities and experiences of the non-disabled body, insights of disabled people are erased (Kafer, 2013, p. 23). This leads to the type of barriers outlined above in terms of both everyday environmentalism and more public forms of organisation and activism.

Therefore, if crip experiences and epistemologies are the object against which capitalist futures are imagined and perceived as desirable, and such capitalist futures are simultaneously the antithesis of sustainable futures, then 'crip experiences and epistemologies should be central to our efforts to counter neoliberalism and access alternative ways of being' (McRuer 2006, pp. 42–43). To do so would be to truly centralise the needs of those that environmental justice movements say that they represent (Johnson, 2011). As Kafer (2013, p. 3) stresses, '[i]n imagining more accessible futures, I am yearning for an elsewhere - and perhaps an "elsewhen" - in which disability is understood otherwise: as political, as valuable, as integral'.

What does cripping environmentalism look like in practice?

An example of centralising crip experiences is provided in the 'Gear Change' report of the Department for Transport (2020a), which provides a vision for increasing walking and cycling in the UK in response to environmental and health challenges and inequalities, including air pollution, climate change and congestion. The Gear Change report is supported by Local Transport Note (LTN) 1/20 (DfT 2020b), which provides national guidance on best practice for cycle infrastructure design to local authorities. What makes Gear Change and LTN 1/20 particularly visionary in terms of disabled environmentalism and disabled cycling is that disabled people's needs are centralised within the development of cycle infrastructure design. LTN 1/20 asserts that infrastructure must be designed for users of all abilities and dis-abilities and recognises the wide range of non-standard cycles – 'cycle resources must be accessible to recumbents, trikes, hand cycles, and other cycles used by disabled cyclists' (DfT 2020b, p. 42) – that can be utilised by disabled people to suit their diverse mobility needs. Access to funding is predicated upon local author-ities adhering to the inclusive design criteria of LTN 1/20, meaning that the needs of disabled cyclists are centralised within the planning and design of cycle infrastructures.

Highlighting policy produced under a Conservative government as an example of centralising disabled people in infrastructural development may seem surprising. Particularly so, where a decade of austerity in the UK has dispropor-tionately negatively impacted disabled people and their families (EHRC, 2017). However Gear Change and LTN 1/20 are important because of the central pos-ition of the role of disabled people within their development, with particular influence from Wheels for Wellbeing. Wheels for Wellbeing is a London-based charity whose work is informed by the experiences of disabled staff, trustees and volunteers, which recognises the physical, emotional and social benefits of cyc-ling and has a vision to get more disabled people cycling for transport, leisure and exercise. Bringing a new and powerful angle to advocacy by applying disability equality legislation and the social model to cycling and supported by cycling advocacy groups who recognise that inclusive and accessible cycling infrastruc-ture benefits all cyclists, Wheels for Wellbeing has successfully worked with national partners, including the Department for Transport to influence attitudes, policy and standards.

Gear Change and LTN 1/20 and the centralised position of disabled bodies in their development are also important for demonstrating the cross-benefits of adopting the language of disabled environmentalism, cripping cycling infrastruc-ture. Centralising disabled bodies within the design of cycling infrastructure is by no means limited to designing for the dimensions of non-standard cycles – such as recumbent trikes – and taking seriously the need for segregation, good surfaces and ensuring correct cambers. And while such designs support disabled people who can and want to cycle, they also provide a safer cycling environ-ment for everyone, including women and children who are less likely to cycle on roads without cycle infrastructures (Aldred and Dales, 2017). Cycle infrastruc-ture designed around the disabled body then not only facilitates environmentalism through supporting disabled people to cycle, but is more broadly associated with the type of infrastructures required for a modal shift away from private car use and towards active travel.

Towards a research and policy agenda in academia

This chapter has highlighted the importance of 'cripping' or centralising the disabled body within environmental justice movements in order to develop a disabled environmentalism disentangled from its normative roots and, thus, capable of imagining truly socially just environmental futures. However, it is very clear that the research agenda on environmentalism and disability requires much broader trans- and interdisciplinary research (Kosanic et al., 2019a; United Nations, 2018; Priebe et al., 2020), both in the Global North and in the Global South. For those working on environment, the research agenda is to recognise the need to make more space for disability studies scholars in our work, to document the effect of policies on disabled people and to understand how disabled people will be affected by the recommendations that we make. We also believe that there is space for disability scholars to investigate environmental policy and practice, welcoming both trans- and interdisciplinary engagement from that field.

Considering, however, that academia is a space in which bodies of knowledge are 'debated, determined, taught, examined and perpetuated' (Gillberg, 2020, p. 12), academia has inevitably had a critical role in the minimal presence of disability in the environmental agenda and environmental justice movements (Johnson, 2011; Kafer, 2013). This is unsurprising considering that academia, particularly in its increasingly neoliberal form, is premised upon non-disabled membership and is therefore endemic with ableism (Brown, 2020). While such ableism is not necessarily rooted within academia itself, but more broadly, and as previously discussed, historically rooted within perceptions of the inferior body, it is further exacerbated by processes of marketisation. Those within academia who cannot achieve the highest level of productivity and efficiency, disabled or otherwise, are pushed out (Brown and Leigh, 2018; Brown, 2020). This is because 'the only way in which neoliberal academia enables subjects to be part of the academic body is by being high-performing and marketisable' (Peruzza, 2020, p. 41), while those considered 'not calculable enough' are cast out (Peruzza, 2020, p. 41). Subsequently, disabled, chronically ill and neurodivergent people are unrepresented within academic staff (Brown and Leigh, 2018) where, despite 16% of the working-age UK population being disabled, only 4.5% of academic staff have declared a disability or impairment (HESA, 2019). Not only is this a social injustice, but it is also a loss to academia and the development of critical theory surrounding all academic works, including environmentalism, particularly as lived experience can be central to the tenets of its development. This is not to advocate for funnelling disabled academics towards disability studies or the integration of disability studies within their work, which too would be a social injustice if not their desired or inspired field to work within (Brown and Leigh, 2018), but rather to highlight that theory and practice produced within an ableist academy will be inherently ableist in scope.

If we are to develop an environmental justice movement that truly responds to all people's needs (Johnson 2011) and leaves no one behind (Leach et al., 2018; UN, 2018), then we not only need to crip environmentalism but also to consider the wider institutions implicated within such movements and the roles and responsibilities – and maybe even power? – of those within these institutions.

References

Adams, M., Khan, N., Begum, S., Wirz, S., Hesketh, T. and Pring, T. (2012) 'Feeding difficulties in children with cerebral palsy: Lost-cost caregiver training in Dhaka, Bangladesh', *Child: Care, Health and Development,* vol 38, no 6, pp. 878–888.

Ahmed, S. (2010) *The Promise of Happiness.* Duke University Press, Durham.

Ahmed, S. (2017) *Living a Feminist Life.* Duke University Press, Durham.

Alaimo, S. (2010) *Bodily Natures: Science, Environment, and the Material Self.* Indiana University Press, Bloomington, IN.

Aldred, R. and Dales, J. (2017) 'Diversifying and normalising cycling in London, UK: An exploratory study on the influence of infrastructure', *Journal of Transport and Health,* vol 4, pp. 348–362.

Baker, S., Reeve, M., Marella M., Roubin, M.D., Caleb, N. and Brown. T. (2017) 'Experiences of people with disabilities during and after Tropical Cyclone Pam and recommendations for humanitarian leaders', *Proceedings of 1st Asia Pacific Humanitarian Leadership Conference,* Melbourne, Australia.

Bell, K. (2014) *Achieving Environmental Justice.* Policy Press, Bristol.

Bell, K. (2020) *Working-Class Environmentalism.* Palgrave Macmillan, London.

Bell, S.L., Tabe, T. and Bell, S. (2020) 'Seeking a disability lens within climate change migration discourses, policies and practices', *Disability and Society,* vol 35, no 4, pp. 1–6.

Beniston, M. and Stephenson, D.B. (2004) 'Extreme climatic events and their evolution under changing climatic conditions', *Global and Planetary Change,* vol 44, no 1–4, pp. 1–9.

Brondizio, E.S., O'Brien, K., Bai, X., Biermann, F., Steffen, W., Berkhout, F., Cudennec, C., Lemos, M.C., Wolfe, A., Palma-Oliveira, J. and Chen, C.T. (2016) 'Re-conceptualizing the Anthropocene: A call for collaboration', *Global Environmental Change,* vol 39, pp. 318–327.

Brooks, S.B. and Snelling, T. (2018) 'Jimmy's resistance, or killing the joy of cruel optimism in South Park', in J.L. Schatz and G.A. Jefferson (eds.), *The Image of Disability: Essays on Media Representations.* McFarland & Company, Jefferson. NC.

Brown, N. and Leigh, J. (2018) 'Ableism in academia: Where are the disabled and ill academics?', *Disability and Society,* vol 33, no 6, pp. 985–989.

Brown, N. (2020) 'Introduction: Theorising ableism in academia', in N. Brown and J. Leigh (eds.), *Ableism in Academia: Theorising Experiences of Disabilities and Chronic Illnesses in Higher Education.* UCL Press, London.

Büchs, M. and Schnepf, S.V. (2013) 'Who emits most? Associations between socio-economic factors and UK households' home energy, transport, indirect and total CO_2 emissions', *Ecological Economics,* vol 90, pp. 114–123.

Chahal, K. (2004) 'Experiencing ethnicity: Discrimination and service provision', Joseph Roundtree Foundation, York.

Chapman, L. (2020) 'Still getting away with murder: Disability hate crime in England', Inclusion London, London.

Charles, A. and Thomas, H. (2007) 'Deafness and disability—Forgotten components of environmental justice: Illustrated by the case of Local Agenda 21 in South Wales', *Local Environment,* vol 12, no 3, pp. 209–221.

Department for Transport (2020a) 'Gear change: A bold vision for cycling and walking', DfT, London.

Department for Transport. (2020b) 'Cycle infrastructure design', Local Transport Note 1/20, DfT, London.

Díaz, S. et al. (2018) 'Assessing nature's contributions to people', *Science,* vol 359, no 6373, pp. 270–272.

Díaz, S. et al. (2019) 'Pervasive human-driven decline of life on Earth points to the need for transformative change', *Science*, vol 366, no 6471, eaax3100.

Díaz, S. et al. (2020) 'Set ambitious goals for biodiversity and sustainability', *Science*, vol 370, no 6515, pp. 411–413.

Di Chiro, G. (2010) 'Polluted politics? Confronting toxic discourse, sex panic, and eco-normativity', in C. Mortimer-Sandilands and B. Erickson (eds.), *Queer Ecologies: Sex, Nature, Politics, Desire*. Indiana Press, Bloomington, IN.

Edelman, L. (2004) *No Future: Queer Theory and the Death Drive*. Duke University Press, Durham.

Equality and Human Rights Commission (EHRC) (2017) 'Being disabled in Britain. A journey less equal', EHRC, London.

Fenney, D. (2017) 'Ableism and Disablism in the UK Environmental Movement' *Environmental Values*, White Horse Press, vol. 26, no 4, pp. 503-522.

Fenney, D. and Snell, C. (2011) 'Exceptions to the green rule? A literature investigation into the overlaps between the academic and UK policy fields of disability and the environment', *Local Environment*, vol 16, no 3, pp. 251–264.

Fenney Salkeld, D. (2016) 'Sustainable lifestyles for all? Disability equality, sustainability and the limitations of current UK policy', *Disability and Society*, vol 31, no 4, pp. 447–464.

Fenney Salkeld, D. (2019) 'Environmental citizenship and disability equality: The need for an inclusive approach', *Environmental Politics*, vol 28, no 7, pp. 1–22.

Gillberg, C. (2020) 'The significance of crashing past gatekeepers of knowledge: Towards full participation of disabled scholars in ableist academic structures', in N. Brown and J. Leigh (eds.), *Ableism in Academia: Theorising Experiences of Disabilities and Chronic Illnesses in Higher Education*. UCL Press, London.

Ginsburg, F. and Rapp, R. (2017) 'Cripping the new normal: Making disability count', *Alter*, vol 11, no 3, pp. 179–192.

Goodman, N. (1978) *Ways of Worldmaking*. Hackett Publishing, Indianapolis, IN.

Hawkins, R. (2012) 'Shopping to save lives: Gender and environment theories meet ethical consumption', *Geoforum*, vol 43, no 4, pp. 750–759.

Hemingway, L. and Priestley, M. (2006) 'Natural hazards, human vulnerability and disabling societies: A disaster for disabled people?', *The Review of Disability Studies: An International Journal*, vol 2, no 3, pp. 57–67.

Higher Education Statistics Agency (HESA) (2019) 'Higher Education Staff Data', HESA, London.

Imrie, R. (1996) *Disability and the City: International Perspectives*. Paul Chapman, London.

Imrie, R. and Thomas, H. (2008) 'The interrelationships between environment and disability', *Local Environment*, vol 13, no 6, pp. 477–483.

IPBES. (2019) 'Global assessment report on biodiversity and ecosystem services of the Intergovernmental Science-Policy Platform on Biodiversity and Ecosystem Services', https://ipbes.net/global-assessment.

IPCC (2018) 'Summary for policymakers', *Global Warming of 1.5°C*. www.ipcc.ch/sr15/chapter/spm/.

Johnson, V.A. (2011) 'Bringing together feminist disability studies and environmental justice', *Barbara Faye Waxman Fiduccia Papers on Women and Girls with Disabilities*, Centre for Women Policy Studies, www.peacewomen.org/assets/file/Resources/Academic/bringingtogetherfeministdisabilitystudiesandenvironmentaljustice_valerieannjohnso.pdf.

Johnston, J. (2008) 'The citizen-consumer hybrid: Ideological tensions and the case of Whole Foods Market', *Theory and Society*, vol 37, no 3, pp. 229–270.

Kafer, A. (2013) *Feminist Queer Crip*. Indiana University Press, Bloomington, IN.

Kerac, M., Postels, D., Mallewa, M., Jalloh, A.A., Vokuijl, W.P., Groce, N., Gladstone, M. and Molyneux, E. (2014) 'The interaction of malnutrition and neurclogic disability in Africa', *Seminars in Pediatric Neuology,* vol 21, no 1, pp. 42–49.

Klein, N. (2014) *This Changes Everything: Capitalism vs. the Clirate.* Simon & Schuster, New York, NY.

Kosanic, A., Petzold, J., Dunham, A. and Razanajatovo, M. (2019a) 'Climate concerns and the disabled community', *Science,* vol 366, pp. 698–699.

Kosanic, A., Kavcic, I., van Kleunen, M. and Harrison, S. (2019b) Climate change and climate change velocity analysis across Germany', *Scientific Reports* vol 9, no 2196, pp. 1–8.

Kosanic, A. and Petzold, J. (2020) 'A systematic review of cultural ecosystem services and human wellbeing', *Ecosystem Services,* vol 45, pp. 101–168.

Leach, M., Reyers, B., Bai, X., Brondizio, E., Cook, C. et al. (2C18) 'Equity and sustainability in the Anthropocene: A social ecological systems perspective on their intertwined futures', *Global Sustainability,* vol 1, p. e13.

Leipoldt, E. (2006) 'Disability experience: A contribution from the margins towards a sustainable future', *Journal of Future Studies,* vol 10, no 3, pp. 15–32

Liu, S., Costanza, R., Farber, S. and Troy, A. (2010) 'Valuing ecosystem services', *Annals of the New York Academy of Sciences,* vol 1185, pp. 54–78.

Lunga, W., Bongo, P.P., van Niekerk, D. and Musarurwa, C. (2019) 'Disability and disaster risk reduction as an incongruent matrix: Lessons from rural Zimbabwe', *Jamba: Journal of Disaster Risk Studies,* vol 11, no 1, pp. 648–655.

MacGregor, S. (2016) 'Go ask "Gladys": Why gender matters in sustainable consumption research', *Discover Society,* 5 January. https://discoversociety.org/2016/01/05/go-ask-gladys-why-gender-matters-in-energy-consumption-research/.

Magdoff, F. and Foster, J.B. (2011) *What Every Environmentalist Needs to Know about Capitalism.* New York University Press, New York and London.

Martín-López, B., Iniesta-Arandia, I., Garcia-Llorente, M., Palomo, I., Casado-Arzuaga, I. et al. (2012) 'Uncovering ecosystem service bundles through social preferences', *PLoS One,* vol 7, p. e38970.

McRuer, R. (2006) *Crip Theory. Cultural Signs of Queerness and Disab'ity.* New York University Press, New York and London.

MEA (2005) *Ecosystems and Human Wellbeing: Biodiversity Synthesis.* World Resources Institute, Washington, D.C.

Middlemiss, L (2018). *Sustainable Consumption: Key Issues.* Routledge, Abingdon.

Mildenberger, M. (2019) The Tragedy of the *Tragedy of the commons.* Scientific American. https://blogs.scientificamerican.com/voices/the-tragedy–of-the-tragedy-of-the-commons/.

Morris, Z., Hayward, R. and Otero, Y. (2018) 'The political determinants of disaster risk: Assessing the unfolding aftermath of Hurricane Maria for people with disabilities in Puerto Rico', *Environmental Justice,* vol 11, no 2, pp. 89–95.

Oliver, M. (1990) *The Politics of Disablement.* Macmillan, London.

Omolo, N. and Mafongoya, P.L. (2019) 'Gender, social capital and adaptive capacity to climate variability: A case of pastoralists in arid and semi-arid regions in Kenya', *International Journal of Climate Change Strategies and Management,* vol 11, no 5, pp. 744–758.

Ourbak, T. and Magnan, A.K. (2018) 'The Paris Agreement and climate change negotiations: Small Islands, big players', *Regional Environmental Change,* vol 18, no 8, pp. 2201–2207.

Oswald, Y., Owen, A. and Steinberger, J.K. (2020) 'Large inequality in international and intranational energy footprints between income groups and across consumption categories', *Nature Energy,* vol 5, no 3, pp. 231–239.

Parmesan, C. and Yohe, G. (2003) 'A globally coherent fingerprint of climate change impacts across natural systems', *Nature,* vol 421, pp. 37–42.

Pascual, U., Balvanera. P., Diaz, S., Pataki, G., Roth, E. et al. (2017) 'Valuing nature's contributions to people: The IPBES approach', *Current Opinion in Environmental Sustainability,* vol 26–27, pp. 7–16.

Peruzzo, F. (2020) 'I am not disabled: Difference, ethics, critique and refusal of neoliberal academic selves', in N. Brown and J. Leigh (eds.), *Ableism in Academia: Theorising Experiences of Disabilities and Chronic Illnesses in Higher Education.* UCL Press, London.

Petzold, Jan, and Alexandre K. Magnan, 2019. Climate change: thinking small islands beyond Small Island Developing States (SIDS). Climatic Change 152(1):145-165

Priebe, J., Mårald, E. and Nordin, A. (2020) 'Narrow pasts and futures: How frames of sustainability transformation limit societal change', *Journal of Environmental Studies and Sciences,* https://doi.org/10.1007/s13412-020-00636-3

Pring, J. (2018) 'Police force admits passing footage of disabled protestors to DWP', *Disability News Service,* 20 December. www.disabilitynewsservice.com/police-force-admits-passing-footage-of-disabled-protesters-to-dwp/.

Pring, J. (2019) 'Police force admits agreement to share information about protestors with DWP', *Disability News Service,* 25 July. www.disabilitynewsservice.com/police-force-admits-agreement-to-share-information-about-protesters-with-dwp/.

Rahim, Z. (2018) 'Police force admits passing disabled anti-fracking protestors' details to DWP', *The Independent,* 24 December. www.independent.co.uk/news/uk/home-news/police-disabled-protesters-fracking-blackpool-lancashire-dwp-fraud-a8696381.html.

Ray, S.J. (2013) *The Ecological Other: Environmental Exclusion in American Culture.* University of Arizona Press, Tucson, AZ.

Ray, S.J. and Sibara, J. (2017) 'Introduction', in S.J. Ray and J. Sibara (eds.), *Disability Studies and the Environmental Humanities: Toward an Eco-Crip Theory.* University of Nebraska Press, Lincoln and London.

Renke, S. (2020) 'You're probably being ableist and you don't even know it'. *Metro,* 11 February. https://metro.co.uk/2020/02/11/what-does-being-ableist-mean-12221655/.

Scope (2019) 'The disability price tag: Policy report', www.scope.org.uk/campaigns/extra-costs/disability-price-tag/

Smith, F., Simard, M., Twigg, J., Kett, M. and Cole, E. (2017) 'Disability and climate resilience: A literature review', Leonard Cheshire and UKAID. www.ucl.ac.uk/epidemiology-health-care/sites/iehc/files/Disability_and_Climate_Resilience_Lit_review.pdf.

Thomas, C. (1999) *Female Forms: Experiencing and Understanding Disability,* Open University Press, Buckingham.

UN. (2018) 'Realization of the Sustainable Development Goals by, for and with persons with disabilities', *UN Flagship Report on Disability and Development,* pp. 1–369.

Walsh-Warder, M. (2016) 'The disproportionate impact of Hurricane Katrina on people with disabilities', *Verge: The Goucher Journal of Undergraduate Writing,* vol 13, pp. 1–20.

Watts, N., Amann, M., Arnell, M., Ayeb-Karlsson, S., Belesova, K., Berry, H., Bouley, T., Boykoff, M., Byass, P. and Cai, W. (2018) 'The 2018 report of the Lancet Countdown on health and climate change: Shaping the health of nations for centuries to come', *The Lancet,* vol 392, pp. 2479–2514.

Wheels for Wellbeing (2018) 'Assessing the needs and experiences of disabled cyclists – Annual survey', https://wheelsforwellbeing.org.uk/wp-content/uploads/2019/04/Survey-report-FINAL.pdf.

Wheels for Wellbeing (2019) 'A guide to inclusive cycling', https://wheelsforwellbeing.org.uk/wp-content/uploads/2019/06/FINAL.pdf.

Whitmee, S., Haines, A., Beyrer, C., Boltz, F., Capon, A.G., de Souza, B.F., Dias, A., Ezeh, H. et al. (2015) 'Safeguarding human health in the Anthropocene epoch: Report of the Rockefeller Foundation-Lancet Commission on planetary health', *The Lancet Commissions*, vol 386, pp. 1973–2028.

WHO (2011) 'World Report on Disability', Geneva, Switzerland.

Withers, A.J. (2012) 'Disablism within animal advocacy and environmentalism', in A.J. Nocella, J.K.C. Bentley and J.M. Duncan (eds.), *Earth, Animal, and Disability Liberation: The Rise of the Ecoability Movement*. Peter Lang Publishing, New York, NY.

Wilkinson, R. and Pickett, K. (2009) *The Spirit Level: Why More Equal Societies Almost Always Do Better*. Allen Lane, London.

Wingate, D. (2019) 'From environmental Malthusianism to ecological modernisation: Toward a genealogy of sustainability', PhD Thesis, University of Leeds, UK.

Wolbring, G. (2009) 'A culture of neglect: Climate discourse and disabled people', *Journal of Media and Culture,* vol 12, no 4, https://doi.org/10.5204/mcj.173.

Zwiers, F.W., Alexander, L.V., Hegerl, G.C., Knutson, T.R., Kossin, J.P., Naveau, P. et al. (2013) 'Climate extremes: Challenges in estimating and understanding recent changes in the frequency and intensity of extreme climate and weather events', in G.R. Asrar and J.W. Hurrell (eds.), *Climate Science for Serving Society*. Springer, Netherlands.

3

HOW ENVIRONMENTALISM INCLUDES AND EXCLUDES WOMEN

Planning, the personal and the planet

Clara Greed

Introduction: what is the problem?

This chapter investigates how environmentalism has included and excluded women. It discusses the extent to which environmentalism has benefitted or disadvantaged women, particularly in respect of developing gender-sensitive policies. Particular reference is made to the United Kingdom (UK) situation, though with some international comparisons. The ambit of environmentalism is broad, and entangled with other major political, cultural and scientific movements, so that it is impossible to delineate it concisely. It is not a monolithic movement and comprises a wide range of perspectives and components (Wardle et al., 2019; Newell, 2020). In general, though, the environmental movement has often been seen as male dominated, in terms of its origins, objectives, policies and leadership (Venkateswaran, 1995; Parpart, 2015). Granted, some women have clearly contributed to the movement, especially through ecofeminism, and some have been highly influential leaders, campaigners and politicians (Buckingham-Hatfield, 2000). Women have been key in introducing the concept of sustainability, which integrates social and economic considerations alongside environmental factors (Brundtland, 1987; Dempsey et al., 2011; Vallance et al., 2011). But, whilst the movement has been positive in raising global awareness of many valid environmental and ecological issues, it has not always been inclusive (Greed, 2011; Bell, 2020).

In the first part of the chapter, relevant roots of environmentalism are outlined, and its scope and nature are discussed. Throughout, a gender lens will be applied to environmentalism, with particular attention to women's presence or absence in the evolving narrative. Environmentalists have sought to address issues including climate change, depletion of natural resources, pollution and threats to wildlife *inter alia*. This emphasis on 'the planet', along with a rather 'disembodied' people-less approach to policy issues, has had some negative effects (Longhurst, 1997). Arguably, environmentalism has sometimes been insensitive to the needs of human beings, especially the practical bodily needs of women (Greed, 2011, 2016a). It is often set at too high a level, being concerned with broad global issues (Gahrton, 2015). Relatively speaking, women are more concerned about the detailed practical local issues they encounter in their daily lives, especially recycling and sanitation issues (Buckingham et al., 2005; George, 2008). Many women environmentalists have found themselves fighting at the local level against the effects of male-dominated global environmental policies that had made their Lves even worse. For example, restricting local people's access to areas designated as game reserves in the name of conservation may limit their access to food and water (Kevane, 2015).

A particular concern has been the lack of inclusion of the bodily and often 'messy' female issues of pregnancy, menstruation, nappies and breast feeding within the environmental agenda. If such biological functions were mentioned by male environmentalists, it was often to condemn women for depleting natural resources and creating 'waste' by buying and disposing of nappies and menstrual products to deal with these natural bodily functions (Buckingham et al., 2005; Fisher, 2008; Sommer and Sahin, 2012; Jewitt and Riley, 2014; Greed, 2016a). Women feel more blamed than men because they are the main shoppers, consumers and carers, having to travel and use up resources as they undertake household tasks. Recycling and rubbish sorting skills have become an extension of female housework duties, as 'green has become the new pink' (Oates and McDonald, 2006, p. 417). In contrast, the female warriors of the ecofeminist movement have much greater ambitions: to change the very way we live and to include women in all aspects of environmental policy. Therefore, the second part of the chapter explains the scope and nature of ecofeminism, which has sought to highlight all the issues male-dominated environmentalism has left out. Ecofeminism was conceived by the coming together of environmentalism and feminism: two of the great ideological movements of the 20th century (Merchant, 1992, 1996; Mies, 2014; Gaard, 2019).

The emphasis upon the *natural* environment has also resulted in less emphasis, relatively speaking, upon the *built* environment. The latter is important, as over 50% of the world's population are now urbanised (Burdett and Sudjic, 2012). Many environmental campaigners demand a reduction in pollution and congestion, and therefore a reduction in motor-car use – 'instantly' (Berners-Lee, 2019; Extinction Rebellion, 2019). There seems to be little understanding of why many people use cars in the first place, especially the very real time and distance constraints they experience in their daily lives. In the UK the majority of the population live in suburban areas, some distance from where they work, shop and go to school, often with limited public transport options (especially outside London), the result of past town planning policy mistakes, particularly the emphasis upon decentralisation (Greed, 2012). The importance of urban land-use planning policy needs to be recognised by environmental campaigners, as a means of reconfiguring urban land-use patterns, especially the relationship between residential, school, childcare and employment locations, in order to reduce the need to travel by car (Lindkvist and Joelsson, 2019). Whilst many town planners are commendably working to achieve such changes, and to create the new 'green' city, the planning profession is still so male dominated that women's different travel patterns and environmental needs are often ignored (Greed, 1994; Sánchez and Neuman, 2020).

Therefore, in the third part, the 'women and urban planning' movement and its policies will be discussed. It is contended that this movement may be seen as the true, but often ignored, form of environmentalism. Women planners have long argued for 'the city of everyday life', which would embody women's 'different' travel patterns, work and lives, and thus would result in a more sustainable city for all (Greed and Johnson, 2014, chapter 10). To conclude, a toolkit for mainstreaming gender into environmental policy and leadership is presented, adapted from previous work on identifying and integrating women's needs into urban planning.

Casting a gender lens on the history of environmentalism

Though there have been environmental struggles around the world for centuries, the word 'environment', and the concept of 'environmentalism', first emerged in the 1960s (Wardle et al., 2019). Before environmentalism was given a name, there was a range of emerging groups in the Global North concerned with 'saving the planet', often linked to a pessimistic form of futurology (Toffler, 1973). Two strands can be identified, a concern with economic factors and a focus on ecological issues. The emerging environmental crusade did not exist in a vacuum but its roots were intertwined with a range of other movements and campaigns.

Economic concerns regarding what was seen as 'overpopulation', especially in the Global South, was the source of alarm regarding the depletion of natural resources and predictions of global starvation. The Club of Rome, which mainly consisted of wealthy American male Western economists, businessmen and academics (though Donella Meadows was a lead author), published the *Limits to Growth*. This publication argued that the earth could not sustain continuing high levels of growth (Meadows et al., 1972). The Sierra Club founded in 1960 had similar concerns (Gahrton, 2015).

From a feminist perspective, it was clear that 'women', especially those in the Global South, were seen to be particularly guilty of causing overpopulation (all virgin births?). Even in the UK, as I well remember from the late 1970s and 1980s, feminist campaigns for better childcare provision which would enable women to work and have careers were likely to be dismissed as selfish in 'encouraging women to breed' (Rothman and Beech, 1987; WEB, 1987; Greed, 1994). Early environmentalism was closely tied to both birth control and eugenics. The need to control population, although ostensibly economic in motivation, undoubtedly had eugenic and Malthusian tones (Malthus, 1798). Demand for population control continues today. The people who make these judgments still mainly consist of elite, Western males (usually with several children), who may be seen as having a rather colonial perspective towards the people of the world, 'who are not like them'. For example, David Attenborough, the famous naturalist and television documentary maker, has stated, '[a]ll of our environmental problems become easier to solve with fewer people' and established 'Population Matters' a campaign to reduce population growth especially in African countries (Attenborough, 2020). In contrast, women have argued that, as women gain education, equality and control over their own lives and families, population will level out naturally, without imposing draconian, top-down, male population controls.

Whilst the early environmental groups were mainly uncritical of capitalism, there were also some that were more socially motivated, including socialist groups. These were concerned with global inequalities, exploitation of cheap labour in the developing world, global health issues and poverty. According to these groups, rather than seeing the people, especially women in the Global South, as 'the problem', international capitalism was seen as the reason for the exploitation of the planet's natural resources and its people (Klein, 2015). So, there were overlaps between socialism and environmentalism. But, ironically, classical Soviet Marxism (like Western capitalism) had promoted industrialisation as the revolutionary way towards social and economic advancement. Mining, heavy industry and nuclear power stations were all seen as signs of progress, and the resultant pollution, and the depletion of natural resources, did not figure strongly in the debate.

There was always an uneasy relationship between socialism and feminism, not least because of the low value put on women's work with greater emphasis being put on heavy manual labour and other male, macho occupations. Heidi Hartmann summed this up as 'the unhappy marriage of Marxism and Feminism' (Hartman, 1981, pp. 1–42). Women's traditional, and still essential work, such as childcare and home-making, was likely to be seen as trivial, selfish, bourgeois and secondary, detracting from the 'real' [male] issues (Attwood, 2010). A similar fraught relationship may be seen to exist between environmentalism and feminism (Salleh, 1997). Environmentalism has arguably taken the spotlight away from unresolved gender-related policy issues. For example, as a town planner, I have often been told by male policymakers, 'oh we've done women, you should be concerned with the environment', as if the two were separate. In fact, they are strongly interrelated, and gender interacts with every aspect of policy-making (Reeves, 2005; Crenshaw, 2019).

The second strand of environmentalism was concerned with 'man' 'saving nature' and 'saving the planet'. In the 19th century, an obsession with nature and a reaction to the Industrial Revolution may be found in the Romantic poets (Wordsworth), in paintings (Constable) and in a desire to explore the countryside (Buckingham Hatfield, 2000). But human beings, especially women, were only likely to be included as picturesque 'figures in the landscape', not as vital agricultural workers with their own needs and lives (Davidoff et al., 1976). A new interest in mastering, controlling and classifying nature was evident (Plumwood, 1993). But women were generally left out of this construct, being neither 'man' nor 'nature'. As Susan Buckingham has explained (Buckingham-Hatfield, 2000), 'women' had been seen, from ancient times, in classical Western philosophy and religion as not being part of 'mankind' but as part of 'nature', something that had to be controlled and part of the 'problem'. They were not seen as human beings in their own right, with their own environmental, social and economic problems.

Policies concerned with conserving wildlife and preventing species extinction, especially in Africa, often prioritised animals over humans (Adams, 2014). Women's lives were made more difficult by restrictive controls on local people's access to their ancestral lands in the name of rural conservation. Local women's 'domestic' roles as gatherers of fire wood and water, and their traditional land rights as cultivators of small plots for food production, were 'invisible' to mainly male expat agricultural advisors and wildlife campaigners (Kevane, 2015). Foreign aid programmes tended to concentrate on the needs of male farmers (giving them tractors and other male technology). They ignored the existence of small-scale female farmers and market traders that kept many countries out of starvation, and all the so-called housework they undertook to feed and care for their families, which might involve walking 5 miles a day just to collect water (Albuquerque and Roaf, 2012).

In the latter part of the 20th century, a more scientific approach to global environmentalism developed. Nature had replaced God as the 'First Cause' and Creator, as evolutionary theory supplanted church teaching and 'man' (not God) became the Master of the universe. A whole range of research-based scientific activities gained prominence, including monitoring climate change and species extinction, and observing 'the hole in the ozone layer', revealed by both terrestrial field research and new satellite data. This scientific approach to environmentalism often seemed even more people-less than earlier versions of the movement. Science was still a male-dominated field and even today there are far fewer women than men scientists. There was limited opportunity for women to cast a gender lens over scientific proof or truth or to ask 'different' questions about the environment (Merchant, 1983).

But the situation is complex in that environmentalism contains many other strands. There was also a growing social and humanitarian component. The organisations involved might not actually see themselves as being environmentalists as such, though they too were dealing with the effects of environmental decline. For example, there were many humanitarian aid agencies and emergency health aid bodies (Oxfam, MSF), as well as long-established church and missionary organisations, dealing with the needs of human beings experiencing problems of health, poverty, lack of education, famine and limited water supply in the Global South – all exacerbated by environmental change. Many women were involved in such organisations as aid workers, volunteers, health professionals, educators, nuns and missionaries. But many were marginal or invisible to the dominant male environmental movement. Nowadays, women aid workers and health professionals are more numerous, and arguably women have been given greater recognition as researchers and activists, within a more feminised development and health agenda (Melies et al., 2011; Parker and Sommer, 2012; Coles et al., 2015; Datta et al., 2020).

As environmentalism progressed, and the related 'green' movement developed, one begins to see a more sensitive, social, spiritual emphasis, as the Age of Aquarius dawned. For example, Schumacher's book, *Small is Beautiful* (Schumacher, 1973) gave more credence to the importance of small local communities in the Global South. Grassroots movements such as Friends of the Earth (FOE) initially appeared more inclusive but subsequently were male dominated, and by a particular type of male at that (Wardle et al, 2019). On the one hand, the environmental movement counted amongst its promoters traditional capitalists (who foresaw the profit to be had from green environmental products) and aristocracy, such as Prince Charles and Baronet Jonathan Porritt, one of the former leaders of the Green Party. But one also sees rather a mystical, alternative, and often somewhat hippy, version of environmentalism developing, as promoted at Glastonbury. It spawned a particular type of male – a lone unkempt, left-wing, activist such as 'Swampy'. All these are very different images of what an environmentalist 'looks like' – but all are male.

Where were the women?

Whilst the majority of leaders of the early environmental movement in the Global North were male, there were individual, significant women which contributed to its development. For example, Octavia Hill had been highly influential in Victorian and Edwardian society, not only as a housing reformer, but as an early campaigner for the protection of the countryside from the ravages of the Industrial Revolution (Darley, 1990). She was a key figure in the founding of the National Trust. This early form of 'environmentalism' was not yet global or planetary in scope but concerned with the British landscape, countryside, farmland, and flora and fauna. Botany was one of the favourite subjects studied by Victorian ladies, although such pre-occupations were not necessarily seen as 'real science' but as yet another example of limited bourgeois female horizons. Rachel Carson in North America (Breton, 1998) was a key post-war figure in identifying in her book, *Silent Spring* (Carson, 1962) the devastating effects of pollution and the destruction of nature by modern industrialised scientific farming methods, especially the use of pesticides. She died young of cancer and seemed to exist at the margins of the male-dominated early environmental movement of the time. Petra Kelly was also an influential figure, especially in the politicisation of environmentalism with the founding of the Green Party in Germany (Parkin, 1994).

Women leaders and politicians also achieved prominence in widening the definition and agenda of environmentalism through generating the concept of sustainability in which environmental, social and economic factors, including gender equality, were linked together at the UN Rio Conference (UN, 1992). Pressure for such an approach had been building since the Brundtland Report *Our Common Future* (Brundtland, 1987). This was instigated by Gro Harlem Brundtland, who was both a previous Norwegian Prime Minister and a head of the World Health Organization (WHO). She was part of a wider Scandinavian and North European feminist movement strongly linked to concerns with both the natural and built environments (Sánchez and Roberts, 2013). Likewise, the Healthy Cities movement, particularly in relation to the work of the WHO, has facilitated a widening of the environmental movement to include concern with the health of human beings as well as the planet (Barton and Tsourou, 2006). One key objective of the sustainability agenda was to ensure that environmental resources are protected and maintained for future generations. From a feminist perspective one could argue that there will be no future generations if women's global concerns regarding health, child-bearing, childcare, access to clean water, and pollution are not addressed as part and parcel of environmentalism. For example, Mary Robinson, past president of Ireland, has argued that climate change was a problem created by men that could only be solved by applying feminist solutions (Robinson, 2018). Some progress is being made in that the UN's Sustainable Development Goals (SDGs) (UN, 2015) do now include, to some extent, women's needs (especially in SDG 5 to 'Achieve gender equality and empower all women and girls' and SDG 6 on sanitation and water), a great improvement on the previous Millennium Development Goals (MDGs) which were less gender-specific (UN, 2000).

Coming up to the present-day, Greta Thunberg has almost gained sainthood status, like Joan of Arc. She is a prominent figure in the environmental movement, particularly amongst young people and followers of Fridays for the Future and Extinction Rebellion, and as a valuable role model in raising environmental awareness amongst school students (Thunberg, 2019). But her speeches and publications have not (yet) manifested a gender awareness of women's and girls' 'different' environmental needs. If one looks across the decades at the leadership characteristics of those involved in environmentalism, the green movement and green politics in the Global North, one could find that most of the leaders are male. As a result, in 1997, Bernadette Vallely set up the Women's Environmental Network (info@wen.org.uk) as an alternative to Green Party policies (Vallely, 1990). But other powerful women, such as Caroline Lucas, a previous leader of the Green Party of England and Wales, have stuck with malestream environmental politics whilst trying to insert a wider feminist perspective (Lucas, 2015). Likewise, certain women continue to be prominent in Friends of the Earth as campaigners promoting women's issues (Hutchins, 2018). It was estimated that around 65% of those who attended Extinction Rebellion (XR) protests in 2019 were women, mainly middle-class, White, from Southern England and with a degree (*Times Newspaper*, 16 July 2020). Women have also featured in XR's leadership, with, for example, Gail Bradbrook being one of its founders.

As the decades sped by, the environmental movement became less radical and more acceptable. Today the words 'environmental', 'sustainable' and 'green' are used very widely as signs of approval and good intentions, both by businesses and governments, and have effectively lost their edge. A world of green consumerism has developed, particularly aimed at women, which has tamed and domesticated environmentalism (Roddick, 2008 [who founded the Body Shop]). A look at most women's magazines will indicate the extent to which the cosmetic, fashion and food advertisements attest to the greenness of their products. Consumer-related environmentalism has arguably distracted attention from a whole range of deeper structural, feminist and women's equality issues.

Ecofeminism – women's 'different' approach to environmentalism

Over the years, many women were involved in the mainstream environmental movement, predominantly as volunteers, local campaigners and community activists, and, occasionally, as leaders and influential politicians. But, overall, there was an increasing dissatisfaction with the nature of policies based on male priorities and concepts of the environment. As more women went into higher education, and as the second wave feminist movement developed, there were many attempts from the 1970s onwards by women researchers and activists to include women's perspectives (and/or feminist perspectives) on environmental issues. For example, there were ecofeminist publications on 'global warming' (Buckingham and Le Masson, 2016). So feminism and environmentalism combined to form ecofeminism. Feminism may be defined as the belief that women should have the same rights and opportunities as men and that a woman's perspective should be taken into account in all aspects of human life – including environmentalism. Francoise D'Eaubonne first coined the phrase 'ecofeminism' in her work *Le Féminisme ou la Mort* (*Feminism or Death*) in 1974, in which she stressed the importance of including women's experience and knowledge in our understanding of the earth and the natural world – as, otherwise, environmentalism was incomplete and doomed (D'Eaubonne, 1974). Women used the term 'eco-feminism' not 'enviro-feminism', as ecology deals with the relationship of organisms between one another and to their physical surroundings. Rather than 'man' being seen as in control of, and saving, the environment, ecology was more concerned with collaboration between humankind, including women, and other species within the natural environment. Some branches of ecofeminism have manifested a concern for animals, not as televisual wildlife, photogenic pets, farm stock, or food resources, but as living, sensate, spiritual beings who share this earth (and our cities) with humans (Narayanan, 2016).

Ecofeminism is not a monolithic or unitary movement and several different versions may be identified. These broadly echo the different attitudes within feminism itself towards women's role in society. Environmentalists themselves have always been divided along political lines too, especially in respect of how to create change. Some adopt a conservative protectionist attitude, others take a more liberal reformist perspective and some others promote a more socialist, even revolutionary, approach to achieving change. Within ecofeminism one can find the whole spectrum ranging from practical environmental policymakers to radical political activists. In addition, some ecofeminists embrace and promote their biological role as mothers, even earth mothers. But other ecofeminists question the traditional gender roles which are assigned to women (men never talk about being 'earth fathers'). Either way, ecofeminists were very wary of finding that their role in environmentalism was 'to clean up the planet', tidying up the mess that men had made and wanted to establish their own agenda instead (Plumwood, 1993; Mies, 2014). Hallen (2001) identified ten different types of ecofeminism. Susan Buckingham (2000) also carefully categorises and summarises different versions (Buckingham-Hatfield, 2000, p. 38, and Box 3.3 on pp. 42–43), including liberal, radical, Marxist, cultural, social, deep ecological, and socialist.

Whilst 'women's issues' featured strongly in the ecofeminist agenda, adherents were concerned with an increasingly broad range of generic issues affecting both men and women, including housing, health, employment, education, social infrastructure and social justice (Gaard and Gruen, 1993; Stein, 2004; Bell, 2014, 2016). Ecofeminists have generally been more concerned with practical, local, physical and biological issues, which are not so prominent in the male version of the movement. Women's involvement often started through campaigns with local issues, such as the cancer caused by chemical pollution at the Love Canal residential area of New York (Krauss, 1993; Bryan, 2003). Sanitation, drainage, pollution, toilet provision and disease control initiatives were all prominent too (Greed, 2003; George, 2008). There was also a strong non-Western component in the evolution of ecofeminism from the Global South, especially the Indian subcontinent. Here, there was an emphasis upon local women's actual experiences of dealing with the effects of environmental denudation, in their work as farmers, and as the ones who had to collect water and fuel, at increasingly longer distances (Mies and Shiva, 1993; Venkateswaran, 1995). Local women in the Global South showed they had potential power as village leaders and change makers, without any need for Western male expat intermediaries (Coles et al., 2015).

Whilst many ecofeminists are concerned with secular policy issues, there is also a spiritual, sacred and religious component of the movement (Radford Ruether, 2005; Adams, 2007; Narayanan, 2014; Greed, 2016b). In India, ecofeminism was directly linked with Hinduism and vegetarianism (Narayanan, 2016). In the West, a distinctively feminist emphasis upon goddess-based religions developed, especially regarding Gaia the goddess of the Earth, the power of women as life givers, and the related concepts of Mother Earth and Mother Nature (Bartowski and Swearingen, 1997). But this was seen by some ecofeminists as too essentialist, reflecting patriarchal views about women as mothers, care givers and all the traditional (and oppressive) cultural roles that went along with this 'natural' and 'religious' division of men and women (Biehl, 1991; Warren, 1997). But Mies and Shiva (1993) argued that such a spiritual approach should not be seen as soft or passive, because it promoted women as powerful leaders and guardians of the planet, and could lead to political change.

When evaluating the different types of ecofeminism, one must retain a critical perspective and consider the extent to which they exclude or include all women. One also needs to consider the extent to which the various branches of ecofeminism have facilitated change to the benefit of women, especially in respect of environmental policy-making. Perhaps patriarchy was only too happy, as with feminism itself, to leave the female adherents of ecofeminism to research and publish their ideas, organise mainly female-attended conferences and even argue amongst themselves, whilst having limited impact upon male-dominated environmentalism (Klein, 2015). In particular, much of the language used in the more religious, mystical and 'earth mother' sectors of ecofeminism is not readily understood, or easily transmitted into the world of male-dominated environmentalism.

It often seemed that women and men were speaking two completely different languages and promoting two entirely different cultures when it came to environmentalism. The mystical and religious component of ecofeminism, whilst a positive attribute, was often unintelligible to male environmentalists who came from scientific, bureaucratic, secular backgrounds and who dealt in 'facts' and physical realities (Greed, 2016b). Also, women's concerns with the so-called private realm, and personal issues of childbirth, housework, menstruation, and with what were often seen as over-localised issues, affecting local villages in the Global South, were a world away from the 'big' global issues and prestigious infrastructural environmental projects that men prioritised. However, under the Local Agenda 21 (a UN programme aimed at embedding sustainability principles in local areas and municipal governments), there was an emphasis upon seeing the local as of global significance (as 'glocal') but women's issues still did not figure strongly in this agenda (Greed and Johnson, 2014, pp. 197–207).

As the years rolled on, environmentalism and the green movement became institutionalised as a result of international UN and EU initiatives (ibid). Environmental policies became embedded in UK law and government policy. They became 'normal' and part of the routine work of the bureaucratic middle-classes, especially in the operation of the UK statutory town and country planning system. The physical aspects of environmentalism fitted easily into planning law and policy. The still male-dominated planning system has always been better at dealing with the physical built environment. But the wider social aspects of creating viable and sustainable cities received far less attention within male-dominated urban planning policy. English town planning law had long deemed anything social, such as provision of childcare facilities, to be *ultra vires* (that is not a land use matter) and there was little representation of women in the decision-making processes related to urban planning policy (Greed, 2005; Greed and Johnson, 2014).

The importance of 'women and planning' as de facto environmentalism

Whilst women have argued for a wider remit of 'social' concerns (aren't all planning issues social?) (Greed, 1994) to be included in environmentalism, they have also argued for a more down to earth, spatial change too. But much of the environmental agenda seems to be floating in a spaceless vacuum (Harvey, 1975). The 'women and planning' movement and also the 'women and geography' movements have highlighted the need for fundamental spatial policy change, at city wide, local neighbourhood and detailed street level to meet both women's needs and, in the process, to create more sustainable cities. Indeed, the 'women and planning' movement may be seen as the true, but marginalised, form of environmentalism.

The old city of man

Many of the problems that women encounter at the city-wide level relate to the old city of man, which was based upon traditional patriarchal approaches to land-use planning and city form (Greed, 1994). Post-War Reconstruction planning (much of which still underlies modern city layout and design) aimed to facilitate the male journey to work, especially by car, and to zone land uses in order to separate home from work and, thus, divide the male/female, public/private and industrial/

domestic realms. Post-war clearance, redevelopment programmes, and the building of decentralised housing estates were undertaken to reduce the imagined chaos and muddle of cities and to provide space for the building of new urban motorways (Greed and Johnson, 2014). Such strategies actually contributed to greater travel distances and therefore congestion, pollution and climate change. This type of town planning did not take into account women's different journeys and work activities. Even today, much transport policy and land-use location policy is based on the simple uninterrupted car-based, male 'journey to work' which now is meant to be done by bicycle. But women undertake more complex journeys, often accompanied by small children and shopping, trip chaining from home to childcare and school drop offs, on to work and back via essential food shopping (Greed, 2012). Public transport routes did not cater for such 'non-standard' travel patterns and stop-offs. Building decentralised out-of-town shopping centres required women to make additional journeys, away from their home area, and often necessitated the use of a car because of poor public transport.

Likewise, at the local neighbourhood level, there was an emphasis upon creating mono land-use areas, 'all housing', for example, without the provision of local shops and other necessities, such as community centres, local childcare facilities, and nearby employment opportunities. Green infrastructure and 'public' open space also needed to be scrutinised which often mainly catered to male team sports and offered few amenities for women (Sinnett et al., 2015). At the micro level of street layout and urban design, everything was dominated by the requirements of the motor car, with little consideration of pedestrian journeys (mainly made by women but invisible as for years the Ministry of Transport did not count journeys under a mile, or those made on foot) (Hass-Klau, 1990). The lack of pedestrian inclusion in street layout and local urban design particularly affected women and children, but also disabled people and older people because of the prevalence of steps, kerbs and dangerous crossings. Typically, pedestrians were expected to cross the new urban roadways by using underpasses, much feared from a personal safety perspective, or footbridges over the road, which was impossible for those who could not manage the steps such as older people, disabled people and anybody with a pushchair, heavy shopping or luggage.

The new city of man

But no sooner had women challenged the principles of the 'old city of man' than they found they had to contend with the 'new green city of man', whose policies, although purported to be environmentally progressive, in fact have created a whole new set of problems for women. There is, still, a lack of gender awareness, and women's different activities and travel patterns are still not fully taken into account.

Cycling is strongly encouraged in the new city of man, but fewer women than men cycle for both road safety and personal safety reasons. Whilst cycling may be a viable option for young, and generally male, commuters, women are often accompanied by babies and small children in their journeys around the city of man (WDS,

2005). This is a particularly English problem as some other European countries, such as the Netherlands, do facilitate such journeys (Lindkvist and Joelsson, 2019).

Some older and disabled people simply cannot cycle. Women in particular are 'time poor' and under much pressure to combine work and childcare duties. Public transport is also promoted, but much of the policy seems to be based upon the situation in London where buses and trains are plentiful. In many parts of the UK, people live mainly in suburban areas where there are few buses, no railway stations and travel by car is the only practical option, especially when commuting distances are lengthy and time constraints are constricting. Women often work 'non-standard' and part-time hours whilst public transport is still geared up to cater for the pre-dominantly male rush hour (Greed, 2012).

Walking is seen not only as a recreational activity, but also as a means of commuting, along with cycling. But women are the ones who already do the most walking, and it is their main form of transport to get around the neighbourhood from school to work, to the shops and so forth. Many women, disabled and older people are very concerned about being expected to share footpaths with cyclists (with no bells to warn people of their approach), especially male lycra louts who imagine that they have the right of way over pushchairs, toddlers, visually impaired people, older people and family dogs on leads on pavements and footpaths. The prospect of electric scooters is even more terrifying.

Footpath systems on new housing estates may look 'green' but many do not join up to paths outside the estate in question and may be cut across by both roads and cycle paths. Pedestrian access to local centres, schools and the places women work needs to be taken into account, so we do not create a 'green stage set' in new housing estates, in which everything looks environmentally sound but is impractical for those living there. In this scenario, nothing joins up to other key land uses and developments related to women's daily journeys and work outside the home (Greed, 2011, 2012).

The new city of woman

Women town planners have critiqued the separation of work and home, the prevalence of decentralisation, all of which generate the need to travel in the first place. They have argued for the creation of 'the city of everyday life' of short distances, mixed land uses, multiple centres, with proximity of home and work locations, which are also child-friendly and safe, and take women's as well as men's journey patterns and work into account. In particular, women's caring-related journeys need more attention (Sánchez and Zucchini, 2019). Such an approach would cut down travel and create a more sustainable city. Likewise, there are many aspects of housing estate design, regarding creating accessible local environments, that would need to change. Women planners and architects have long investigated every aspect of urban planning, street design and building design, and conceptualised alternatives, a topic area which is too vast to include in this chapter (but see, e.g., Stimpson et al., 1981; Greed, 1994; Sanchez and Roberts, 2013).

To achieve the sustainable city for all, attention needs to be given to detailed bodily considerations. If the government wants us to leave our cars at home, and cycle, walk and use public transport, then public toilet provision is the missing link in creating sustainable cities. If we use these sustainable alternatives, we cannot simply drive to the nearest motorway service station to relieve ourselves. Toilets are important for women as we need to go more often (because of menstruation, pregnancy, menopause and incontinence). Furthermore, small children, some older people and some disabled people also need adequate toilet provision. They should always be provided at transport termini and local urban centres. Even if toilets are provided, they may only be for men because much of the 'sanitation and sustainability' agenda is still prone to dealing with genderless generalities, or takes the male as default (Ramster et al., 2018; Greed, 2019). This is an international problem because the lack of provision reduces the chances of women being able to travel, work, and go to school, and thus undermines the achievement of key development goals especially in relation to employment, economic development, equality and education (Greed, 2016a).

Conclusion: mainstreaming gender into all aspects of policy-making and leadership

Many environmental problems cannot be solved at the personal level, by provoking people to feel guilty, or just by making a lifestyle change, such as reducing personal use of plastic bags. Most people have little power and cannot, for example, build a new railway station or change road design. Therefore, the women and planning movement has also put an emphasis on 'who' is in charge of urban policy-making and who has created the problems in the first place, as well as looking at 'what' the policies are. There has been a great deal of research and publication promoting women's different policies, but implementation has been very limited. Therefore, in order to integrate gender-aware policies into both town planning and environmental policy it is important to break down the decision-making processes into stages as demonstrated in our gender mainstreaming toolkit (RTPI, 2003). This toolkit is intended to raise awareness to enable planners to take into account the needs of women at each stage of policy-making, as shown in the following set of steps that may be applied to environmental policy-making too (NERC, 2019).

Summary of the Gender Mainstreaming Toolkit applied to environmental policy

Ask these questions in relation to each stage of plan-making and policy development:

- What resources, training and experience are available in the policy-making body on women's perspective and needs?
- How is the policy team chosen? Is it representative of both men and women?
- Who is perceived to be the population under consideration?

- How are the statistics gathered? Are they disaggregated by gender?
- What are the key priorities, values and objectives of the plan?
- Who is consulted and who is involved in public participation?
- How is the plan evaluated?
- Are complaints and pre-existing concerns acknowledged?
- How are the policies implemented and managed?
- How far into the future are the policies and plans monitored and supported?

If these stages are integrated into the plan-making and policy development process, it should lead to consciousness raising amongst the predominantly male apparatchiks of the environmental system. But it is not necessary to believe in the importance of women's issues to use this Toolkit model, or to be female, because it provides planners with something 'concrete' and reasonably doable in logical sequential stages that might be readily fitted into many planning processes.

References

Adams, C. (2007 [1994]) *Ecofeminism and the Sacred*. Continuum, London.
Adams, C. (2014) *Ecofeminism: Feminist Intersections with Other Animals and the Earth*. Bloomsbury, London.
Albuquerque de, C. and Roaf, V. (2012) *On the Right Track: Good Practices in Realising the Rights to Water and Sanitation*. UNESCO, Paris.
Attenborough, D. (2020) *A Life on Our Planet: My Witness Statement and Vision for the Future*. Penguin, London.
Attwood. L. (2010) *Gender and Housing in Soviet Russia: Private Life in Public Space*. Manchester University Press, Manchester.
Barton, H. and Tsourou, C. (2006) *Healthy Urban Planning*. Spon, Oxford.
Bartowski, J.P. and Swearingen, W.S. (1997) 'God meets Gaia: A case study of environmentalism as implicit religion', *Review of Religious Research*, vol 38, no 4, pp. 308–324.
Bell, K. (2014) *Achieving Environmental Justice*. Policy Press, Bristol.
Bell, K. (2016) 'Bread and roses: A gender perspective on environmental justice and public health', *International Journal of Environmental Research and Public Health*, vol 10, 1005.
Bell, K. (2020) *Working-Class Environmentalism: An Agenda for a Just and Fair Transition to Sustainability*. Palgrave Macmillan, London.
Berners-Lee, M. (2019) *There Is No Planet B: A Handbook of the Make or Break Years*. Cambridge University Press, Cambridge.
Biehl, J. (1991) *Rethinking Ecofeminist Politics*. South End Books, Boston, MA.
Breton, M.J. (1998) *Women Pioneers for the Environment*. Northeastern University, Boston, MA.
Brundtland Report (1987) *Our Common Future, World Commission on Environment and Development*. Oxford University Press, Oxford.
Bryan, N. (2003) *Love Canal*. World Almanac Library, New York.
Buckingham, S., Reeves, D. and Batchelor, A. (2005) 'Wasting women: The environmental justice of including women in municipal waste management', *Local Environment*, vol 10, pp. 427–444.
Buckingham, S. and Le Masson, V. (eds.) (2016) *Understanding Climate Change Through Gender Relations*. Routledge, London.
Buckingham-Hatfield, S. (2000) *Gender and Environment*. Routledge, London.
Burdett, R. and Sudjic, D. (2012) *Living in the Endless City*. Phaidon, London.

Carson, R. (1962) *Silent Spring*. Harmondsworth, London.

Coles, A., Gray, L. and Momsen, J. (eds.) (2015) *The Routledge Handbook of Gender and Development*. Routledge, New York.

Crenshaw, K. (2019) *On Intersectionality: Essential Readings*. New Press, New York.

D'Eaubonne, F. (1974) *Le Féminisme ou la Mort (Feminism or Death)*. Hatchard, Paris.

Darley, G. (1990) *Octavia Hill*. Constable, London.

Datta, A., Hopkins, P., and Johnston, L. et al. (eds.) (2020) *Routledge Handbook of Gender and Feminist Geographies*. Routledge, London.

Davidoff, L., L'Esperance, J. and Newby, H. (1976) 'Landscape with figures: Home and community in English society', *International Journal of Urban and Regional Research*, Special Issue on 'Women and the City', vol 2, no 3, pp. 558–563.

Dempsey, N., Bramley, G., Power, S. and Brown, C. (2011) 'The social dimensions of sustainable development: defining urban social sustainability', *Sustainability and Development*, vol 19, pp. 289–300.

Extinction Rebellion (2019) *This Is Not a Drill: Extinction Rebellion Handbook*. Penguin, London.

Fisher, J. (2008) 'Women in water supply, sanitation and hygiene programmes', *Municipal Engineer*, vol 161, pp. 223–229.

Gaard, G. (2019) *Critical Ecofeminism: Ecocritical Theory and Practice*. Lexington Books, Lanham, MD.

Gaard, G. and Gruen, L. (1993) 'Ecofeminism: towards global justice and planetary health', *Society and Nature*, vol 2, no 1, pp. 1–35.

Gahrton, P. (2015) *Green Parties, Green Future: From Local Groups to the International Stage*. Pluto Press, London.

George, S. (2008) *The Big Necessity: Adventures in the World of Human Waste*. Portobello Press, London.

Greed, C. (1994) *Women and Planning: Creating Gendered Realities*. Routledge, London.

Greed, C. (2003) *Inclusive Urban Design: Public Toilets*. Architectural Press, Oxford.

Greed, C. (2005) 'Overcoming the factors inhibiting the mainstreaming of gender into spatial planning policy in the United Kingdom', *Urban Studies*, vol 42, no 4, pp. 1–31.

Greed, C. (2011) 'Planning for sustainable urban areas or everyday life and inclusion', *Journal of Urban Design and Planning*, Proceedings of the Institution of Civil Engineers, vol 164, June, pp. 107–119.

Greed, C. (2012) 'Planning and transport for the sustainable city or planning for people', *Journal of Urban Design and Planning*, vol 165, June, pp. 219–229.

Greed, C. (2016a) 'Taking women's bodily functions into account in urban planning policy: Public toilets and menstruation', *Town Planning Review*, vol 87, no 5, pp. 505–523.

Greed, C. (2016b) 'Religion and sustainable urban planning: If you can't count it, or won't count it, it doesn't count', *Sustainable Development*, vol 24, pp. 154–162.

Greed, C. (2019) 'Join the queue: Including women's toilet needs in public space', *The Sociological Review*, vol 67, no 4, pp. 908–926.

Greed, C. and Johnson, D. (2014) *Planning in the UK: An Introduction*. Palgrave Macmillan, London.

Hallen, P. (2001) 'Recovering the Wildness in Ecofeminism', *Women's Studies Quarterly*, vol. 29, no. 1/2, pp. 216-233

Hartmann, H. (1981) 'The unhappy marriage of Marxism and feminism', in L. Sargent and H. Hartman H. (eds.), *Women and Revolution*. Pluto Press, London.

Harvey, D. (1975) *Social Justice and the City*. Arnold, London.

Hass-Klau, C. (1990) *The Pedestrian and City Traffic*. Belhaven Press, London.

Hutchins, L. (2018) *Why Women Will Save the Planet*. Zed Books in association with Friends of the Earth, London.

Jewitt, S. and Ryley, H. (2014) 'It's a girl thing, menstruation, school attendance, spatial mobility and wider gender inequalities in Kenya', *Geoforum*, vol 56, pp. 137–147.

Kevane, M. (2015) 'Changing access to land for women in sub-Saharan Africa', in A. Coles, L. Gray and J. Momsen (eds.), *The Routledge Handbook of Gender and Development*. Routledge, New York.

Klein, N. (2015) *This Changes Everything: Capitalism versus the Climate*. Penguin, London.

Krauss, C. (1993) 'Women and toxic waste protest: Race class and gender as resources of resistance', *Qualitative Sociology*, vol 16, pp. 247–262.

Lindkvist, S.T. and Joelsson, T. (eds.) (2019) *Integrating Gender into Transport Planning: From One to Many Tracks*. Palgrave Macmillan, London.

Longhurst, R. (1997) '(Dis)embodied geographies', *Progress in Human Geography*, vol 21, pp. 486–501.

Lucas, C. (2015) *Honourable Friends? Parliament and the Fight for Change*. Portobello Books, London.

Malthus, T. (2008 [1798]) *An Essay on the Principle of Population*. Oxford University Press, Oxford.

Meadows, D.H., Meadows, D.L., Randers, J. and Behrens, W. (1972) *The Limits to Growth*. Universe Books, New York.

Melies, I., Birch, E. L. and Wachter, S.M. (2011) *Women's Health and the World's Cities*. University of Pennsylvania Press, Philadelphia, PA.

Merchant, C. (1983) *The Death of Nature: Women, Ecology and the Scientific Revolution*. Harper and Row, New York.

Merchant, C., (1992) *Radical Ecology: The Search for a Liveable World*. Routledge, London.

Merchant, C. (1996) *Earthcare: Women and the Environment*. Routledge, London.

Mies, M. (2014) *Ecofeminism: Critique, Influence, Change*. Zed Books, London.

Mies, M. and Shiva, V. (1993) *Ecofeminism*. Zed Books, London.

Narayanan, Y. (2014) *Religion, Heritage and the Sustainable City: Hinduism and Urbanisation*. Routledge, London.

Narayanan, Y. (2016) 'Where are the animals in sustainable development? Religion and the case for ethical stewardship in Animal Husbandry', *Sustainable Development*, vol 24, pp. 172–180.

NERC (2019 onwards) A systems approach to sustainable sanitation challenges in urbanising China (SASSI). www.complexurban.com/project/sassi/.

Newell, P. (2020) *Global Green Politics*. Cambridge University Press, Cambridge.

Oates, C. and McDonald, S. (2006) 'Recycling and the domestic division of labour: Is green the new pink?' *Sociology*, vol 30, no 3, pp. 417–433.

Parker, R. and Sommer, M. (2012) *The Routledge Handbook of Global Public Health*. Routledge, New York.

Parkin, S. (1994) *The Life and Death of Petra Kelly*. Rivers Oram Press/Pandora, London.

Parpart, J.L. (2015) 'Men, masculinities and development', in A. Coles, L. Gray and J. Momsen (eds.), *The Routledge Handbook of Gender and Development*. Routledge, New York.

Plumwood, V. (1993) *Ecofeminism and the Mastery of Nature*. Routledge, London.

Ramster, G., Greed, C. and Bichard, J.-A. (2018), 'How inclusion can exclude: The case of public toilet provision for women', *Built Environment*, vol 4, no 1, pp. 52–76.

Reeves, D. (2005) *Planning for Diversity: Planning and Policy in a World of Difference*. Routledge, London.

Radford Ruether, R. (2005) *Integrating Ecofeminism, Globalisation and World Religions*. Rowman and Littlefield Publishers, London.

Robinson, M. (2018) *Climate Justice: A Man-Made Problem with a Feminist Solution.* Bloomsbury, London.

Roddick, A. (2008) 'The revolutionary eccentric', in K. Eberhardt Shelton (ed.), *A Woman's Guide to Saving the World.* The Book Guild, Brighton.

Rothman, L. and Beach, J. (1987) 'Local Initiatives in child-care services', *Women and Environments*, vol 9, no 2.

RTPI (2003) *Gender Mainstreaming Toolkit.* The Royal Town Planning Institute, London (authors Dory Reeves and Clara Greed).

Sallel, A. (1997) *Ecofeminism as Politics: Nature, Marx and the Post-Modern.* Zed Books, New York.

Sánchez de Madariaga, I. and Roberts, M. (eds.) (2013) *Fair Shared Cities.* Ashgate, London.

Sánchez de Madariaga, I. and Neuman, M. (2020) *Engendering Cities: Designing Sustainable Urban Spaces for All.* Routledge, London.

Sánchez de Madariaga, I. and Zuchini, E. (2019) 'Measuring mobilities of care: A challenge for transport agendas', in T. Lindkvist Scholten and T. Joelsson (eds.), *Integrating Gender into Transport Planning: From One to Many Tracks.* Palgrave Macmillan, London.

Schumacher. E.F. (1993 [1973]) *Small Is Beautiful: A Study of Economics as If People Mattered.* Vintage, London.

Sinnett, D., Smith, N., and Burgess, S. (eds.) (2015) *Handbook of Green Infrastructure: Planning, Design and Implementation.* Edward Elgar, London.

Sommer, M. and Sahin, M. (2012) 'Overcoming the taboo: Advancing global agenda for menstrual hygiene management for school girls', *American Journal of Public Health*, vol 103, pp. 1556–169.

Stein, I. (ed.) (2004) *New Perspectives on Environmental Justice: Gender, Sexuality and Activism.* Rutgers University Press, New Brunswick, NJ.

Stimpson, C., Dixler, E., Nelson, M. and Yatrakis, K. (eds.) (1981) *Women and the American City.* University of Chicago Press, Chicago, IL.

Thunberg, G. (2019) *No One Is Too Small, to Make a Difference.* Penguin, London.

Toffler, A. (1973) *Future Shock.* Pan Macmillan, London.

UN (1992) *The Rio Declaration on Environment and Development.* United Nations, New York.

UN (2000) *The Millennium Development Goals.* United Nations, New York.

UN (2015) *Sustainable Development Goals*, SDGS. United Nations, New York.

Vallance, S., Perks, C.H. and Dixon, E. (2011) 'What is social sustainability? A clarification of concepts', *Geoforum*, vol 42, no 3, pp. 137–147.

Vallely, B. (1990) *1001 Ways to Save the Planet.* Penguin, London.

Venkateswaran, S. (1995) *Environment, Development and the Gender Gap.* Sage, London.

Wardle, P., Robin, L. and Sverker S. (2019) *The Environment: A History of the Idea.* Johns Hopkins University Press, Baltimore, MD.

Warren, K. (2018 [1997]) *Ecofeminism: Women, Culture, Nature.* University of Indiana Press, Bloomington, IN.

WEB (1987) 'Women's realm', *WEB Newsletter of Women in the Built Environment*, no 6, July, pp. 1–12.

WDS (2005) *Cycling for Women.* Women's Design Service, London.

4

ENVIRONMENTAL MOVEMENTS IN THE GLOBAL SOUTH

Silpa Satheesh

Introduction

On 1 August 1998, protestors formed a human chain across the polluted stretch of River Periyar, circling the Eloor-Edayar industrial belt in Kerala. The river flowed in many different shades carrying the untreated industrial effluents from the chemical industries situated along its banks. The human chain was staged in an array of giant country boats controlled by the local fisherfolk. The event marked the beginning of an organised environmental movement against industrial pollution in the region led by *Periyar Malineekarana Virudha Samithi* (Periyar Anti-Pollution Campaign, PMVS hereafter). PMVS is a grassroots environmental movement constituted by poor and working-class members, including fish workers, farmworkers and daily wage labourers to fight against the release of toxic industrial effluents into the river. Almost 22 years later, on 22 April 2020, the frontline leaders of PMVS and *Janajagratha* (People's Vigilante) lined up on *Pathalam* regulator-cum-bridge to protest the ongoing pollution in River Periyar, which started flowing again in many colours (Satheesh, 2020b). The strikingly similar events of the protests, organised decades apart, show how industrial pollution continues in the region unabated despite the prolonged fight for environmental justice. PMVS stands as an excellent exemplar of poor and working-class environmental movements in the Global South, which fight against the negative externalities created in the name of development.

Despite the long history of mobilisations that span decades, environmental movements in the Global South rarely appear in the mainstream literature on environmentalism and environmental movements. The omission of such movements has resulted in producing a singular and monolithic characterisation of environmentalism as a middle-class phenomenon grounded in post-materialist value orientations. Combining material and ecological grievances in their protest vocabularies, these poor and working-class movements include social, material and environmental issues in their discourses and actions. In this context, the present chapter traces the rich and vibrant history of environmental movements in India to demonstrate its strange omission from the dominant literature on environmentalism. In that respect, the chapter reaffirms the need to decolonise research on environmental social movements by retelling the numerous exemplars of local and grassroots environmental struggles from the Global South. In so doing, the chapter also challenges the blanket characterisation of environmental movements as New Social Movements (NSM) focusing not on class, but on identities.

Debating the history of environmentalism

The history of modern environmentalism is often traced back to the publication of Rachel Carson's *Silent Spring* in 1962. However, many studies have explained how pinpointing this particular event is problematic, considering the long lineage of indigenous and working-class environmental movements in both the Global North and Global South (Taylor, 1993; Montrie, 2011, 2018). As observed by Grove (2002), the extension of the capitalist 'world system' on a global scale between 1200 and 1788 intensified the exploitation of natural resources. The colonial model of resource exploitation undertaken through trading companies, such as the East India Companies of Portugal, the Netherlands, Britain and France, enforced intensive cash-crop plantation activities in addition to clearing forests for agriculture and shipbuilding (Bandyopadhyay and Shiva, 1988; Guha and Gadgil, 1989; Grove, 2002). Highlighting the responses to environmental degradation in history, Grove observes:

> It is a common fallacy to think that globalization and environmental crises are new phenomena, or products only of the post-World War II world. Global environmental concerns are also often considered to be relatively new. But the story of environmentalist reactions to human-induced ecological changes on a global scale is actually more than three centuries old.
>
> *(Grove 2002, p. 50)*

Though Grove (2002) focuses on state-led legislative responses to environmental degradation here, the responses to the colonial plunder of natural resources were much more vibrant and agentic. The rich history of forest conflicts in India presented in the writings of Ramachandra Guha and Madhav Gadgil stand testimony to the mighty struggles organised by indigenous communities against the alienation of natural resources under colonial rule. In their germinal, *State Forestry, and Social Conflict in British India*, Guha and Gadgil (1989) trace the history of forest conflicts in pre-independent India to the 19th century. Predominantly led by tribal communities, these conflicts fought against the exclusionary forest management policies introduced by the colonial administration that curbed the communities' access to the forest resources (Gadgil and Guha, 1993). Guided only by the imperatives of commercial and strategic utility of natural resources, the colonial administration carried out a fierce onslaught on India's forests (Smythies, 1925; Guha, 1985). The colonial forestry regime redefined the property rights in such a way that it conflicted with the earlier systems of local access, use and control. And such changes in resource patterns initiated extensive conflicts over forest produce across the country. Featuring a variety of conflicts against the colonial forest laws organised by indigenous hunter-gatherer communities (Kumaun movement) and shifting cultivators (Baiga and Saora tribal communities' opposition to the restrictions on *Jhum* cultivation), Guha and Gadgil (1989, p. 153) extrapolate the history of organised environmental resistance to as early as the 19th century. One of the essential aspects of these conflicts was the conflation of material and ecological grievances which drove these movements and how they were featured in the pro-test vocabularies. In doing so, these environmental struggles point out the need to broadly define environmentalism to include environmental justice and livelihood struggles.

Thus, instances of indigenous conflicts in the Global South challenge the extant framings of environmentalism in the mainstream literature on two grounds: (1) environmentalism is not a post-1960 phenomenon led exclusively by the middle-class in the Global North, and (2) material grievances play a crucial role in envir-onmental struggles. Given this backdrop, this chapter explores how the poor and working-class environmental movements in the Global South, both histor-ical and contemporary, shake the dominant narratives surrounding environmental movements as (a) a middle-class phenomenon, (b) having a post-materialist orien-tation and (c) one that is grounded in identities and not class. Building on these critiques, the chapter also establishes that classifying the poor and working-class environmental movements in the Global South as NSMs is problematic.

The environmentalism of the poor

The environmental movement is best understood as an 'envelope' that contains a 'variety of socially and discursively constructed ideologies and actions, theories, and practices' (Dwivedi, 2001, p. 12). Despite the replete presence of environmental struggles in the Global South, it is strange to see how much of this envelope remains unopened in the dominant literature when it comes to the poor and working-class environmental movements in the Global South (Satheesh, 2020a). The proliferation of such struggles for environmental justice in the Global South is evident from the ever-increasing number of cases reported in the Environmental Justice Atlas, an online repository of conflicts surrounding environmental issues (Temper et al., 2015). Popularly referred to as 'environmentalism of the poor', these movements challenge the unequal distribution of environmental burdens on poor and marginalised com-munities (Guha, 2002; Martinez-Alier, 2003). Gadgil and Guha (1994, p. 131), breaking the myth of environmentalism as middle-class phenomena, argue that 'poor countries…poor individuals and poor communities within them, have shown a strong interest in environmental issues'. Combining environmental and social justice grievances, these movements expose how environmental degradation and denial of access rights jeopardise the income and livelihoods of resource-dependent communities (Gadgil and Guha, 1994; Karan, 1994).

Notable instances of such struggles featured in the academic literature include the Chipko movement in the Himalaya (Guha, 2000), the Narmada Bachao Andolan movement (Baviskar, 1995; Nilsen, 2010), the fight to save the Amazon rain forest (Revkin, 2004), the struggles in Niger Delta against the Shell corporation (Osaghae, 1995; Obi, 1997), the fight against Coca-Cola in Plachimada (Bijoy, 2006), the opposition led by artisanal fisherfolk against mechanised trawling (Kurien, 1991) and the movements against Vedanta in Odisha (Shrivastava and Kothari, 2012). A careful analysis of these movements brings forth the underlying distributional inequity these movements sought to challenge and how it pits the powerful against the powerless (Gadgil and Guha, 1994).

The Chipko or 'hug the tree' movement, one of the renowned environmental movements in India, featured a struggle between hill villagers and state forest policies that favoured commercial felling of timber over subsistence use (Guha, 2000; Rangan, 2000). The movement that relied on Gandhian *Satyagraha* succeeded in securing a 15-year ban on commercial tree felling in the Himalayan forests of Uttar Pradesh in 1980 (Shiva and Bandyopadhyay, 1986). Similarly, the struggle organised by the Dongria Kondh, a vulnerable tribal community in Odisha, against the dispossession of traditional and sacred lands by a corporation in the name of bauxite mining highlights the competing valuations surrounding environmental resources and the incommensurability of natural resources (Shrivastava and Kothari, 2012). A landmark Supreme Court verdict that stated that forest clearance for the mining project should be given only after taking the consent of village councils in the region vindicated the anti-mining movement (Seetharaman, 2018). The vigorous fight against Coca-Cola in Plachimada, Kerala, signifies how a village led by Mayilamma, a woman leader from the tribal community, fought against the Coca-Cola's groundwater exploitation and succeeded in ousting the multinational corporation from their lands (Raman, 2005; Bijoy, 2006). Challenging the onslaught of neoliberal globalisation and subsequent privatisation of natural resources, Assies describes the *David vs. Goliath* fight in Cochabamba surrounding water (Assies, 2003). Popularly known as the Cochabamba Water War, the movement against the privatisation of water in Cochabamba city in Bolivia stands as an important victory against corporate globalisation in Latin America (Olivera and Lewis, 2004).

The active framing against the powerful has also led to the victimisation of environmental activists in the Global South (Watts and Vidal, 2017). Citing the report published by Global Witness, *The Guardian* reports that around 212 environmental activists were killed in the last year alone for defending their land and environment (Greenfield and Watts, 2020). The killing of environmentalists is happening at an alarming rate, with 40% from indigenous communities (Global Witness, 2020; Chandran, 2020). Chico Mendes, the Brazilian labour leader and conservationist, was assassinated by local ranchers in 1988 while fighting to save the Amazon Rainforest. The killing of Paulo Paulino Guajajara, a Brazilian indigenous leader, by armed loggers in the Amazon frontier region in 2019 also exposes the repression and violence endured by environmentalists (Cowie, 2019). In 1991, Shankar Guha Niyogi, the leader of Chattisgarh Mines Shramik Sangh (CMSS) who moved beyond conventional trade unionism to combine labour and environmental grievances, was murdered (Lin, 1992). Similarly, Ken Saro-Wiwa, who led a peaceful movement for the environmental and human rights of Ogoni people in the Niger delta, was executed by the Nigerian government in 1995 (Na'Allah, 1998). In 2016, the murder of Lenca leader, Berta Cáceres, who was involved in many land and water struggles in Honduras, sparked an international outcry against the violence and intimidation unleashed against environmentalists (Erdős, 2019; Lakhani, 2020; Méndez, 2018). The long list of frontline environmental activists who have lost lives in Columbia, Brazil, Philippines and Mexico stands testimony to the violent backlash from the extractive sector and economic elite.

Calling out the material and resource deprivations caused by state and corporate-sponsored projects, these movements make the material and social justice orientations of their movements explicit. According to Martinez-Alier (2003), the worldwide movements of the environmentalism of the poor challenge the post-materialist thesis[1] (Inglehart, 1990) that environmental preservation would emerge as a desire only after the material necessities of life have been met. In fact, the prevalence of environmental struggles in poor and developing countries in the Global South exposes the bankruptcy of the claim that environmentalism is a characteristic feature of post-industrial societies (Martinez-Alier, 1997). Moreover, it highlights how such singular categorisations erase and obscure the distinct social and political-economic conditions prevalent in the Global South when compared to the post-industrial North (Dwivedi, 2001).

As demonstrated in the examples above, the environmental movements in the Global South are astute in identifying the targets and mobilizing people and resources to fight against environmental injustices. The grievance interpretations underlying these conflicts, thus, recognise the role of structural factors in engendering environmental inequalities and burdens at individual and community levels. Quite distinct from their middle-class counterparts, the environmentalism of the poor calls out the political economy of development that renders people at the social margins vulnerable to the fallouts of modernisation projects (Shiva, 1991). Analysing the ideological trends in Indian environmentalism, Gadgil and Guha (1994:127) identified three distinct ideological perspectives in the movement: Crusading Gandhians, Appropriate Technologists and Ecological Marxists. In other words, when the mainstream environmental movements shy away from identifying the causes of environmental degradation but narrowly focus on improving the quality of life, the environmental movements in the South expose how the predatory model of development derails the lives and livelihoods of people at the social and economic margins in the name of profits (Bandyopadhyay and Shiva, 1988). Studies unravelling the dynamics of Southern environmental movements clarify the need for a political economy lens to understand how such struggles pit the benefactors of economic development against the people who bear its costs (Shiva, 1991; Dwivedi, 2001). As observed by Bandopadhyay and Shiva (1988, p. 1224), 'The intensity and range of the ecology movements in independent India have kept on increasing as predatory exploitation of natural resources to feed the process of development has gone up in extent and intensity'.

However, the predominance of livelihood issues and unequal distribution of environmental burdens in the movement frames does not imply that all of these movements can be confined to a simple and blanket categorisation. Cautioning such tendencies prevalent in existing research that conceive the distinction between Northern and Southern environmental movements as a binary, Dwivedi (2001, p. 24) elaborates on the multidimensionality of environmental movements and reaffirms that Southern environmental movements are 'as much over meanings and knowledge as over material resources'

Red and green shades of environmentalism

The presence of working-class environmental movements in the Global South challenges the popular notion that class is not a unification factor as far as environmental grievances are concerned. The strong working-class constituency and explicit Marxist orientations of some of these struggles underscore how environmental inequality is a class issue (Nilsen, 2008; Satheesh, 2020c). Critiques have called out how popular framings seek to classify environmental movements in the Global South using the tropes of green and red when the reality stands closer to ideological hybridity and internal contradictions within the movement group (Baviskar, 2005). Though such critiques are extremely important in challenging the tendencies to straitjacket environmental movements in the Global South, it would be premature to dismiss the red shade of green politics ingrained within grassroots environmental movements in India. Nilsen (2008) undertook a Marxian analysis of the conflict over dam-building on the Narmada River in western India and explored how the Narmada movement influenced the trajectory of capitalist development in postcolonial India. It is interesting that this is one of the cases Baviskar (2005) used to critique the claims surrounding the red and green agendas of Indian environmentalism.

The movements in Kerala, a south Indian state known for its unique model of development, communist governments and history of working-class struggles (Parayil, 1996; Isaac and Harilal, 1997; Heller et al., 2007), stand out for their leftist orientations and politics. The *Vayalkili* (farm birds) protest in Keezhattoor (Ameerudheen, 2018), *Kandankali* movement (The New Indian Express, 2017), and the Periyar Anti-Pollution Campaign led by PMVS are a few notable instances where environmental grievances are interpreted in terms of class politics. These movements draw heavily from Marxian ideology, and such affiliations are explicit in their mobilizing strategies and protest lexicons. For example, the PMVS in the Eloor-Edayar industrial belt is a movement where the members have declared affiliations with the state's left movement. Despite the active stand-off with the organized trade union movements within the local industries, the frontline leaders unequivocally point out that their fight is against capitalist exploitation and not against the factory workers (Satheesh, 2020b, 2020c). The deep commitment to the Marxian philosophy and idioms of class-consciousness is emergent in the campaign materials produced by PMVS. A pamphlet announcing a discussion forum on *Marxism, Environment, and Development* published by PMVS in 2008 reads as follows:

> The capitalist forces that plunder soil, water, and all the natural resources to reap profits threaten the survival of human beings and our nature. The state, on the other hand, is invested in devising policies and programs to facilitate the transfer of the title deeds for our natural resources to such capitalistic forces…Though such blatant conquests on nature or state-capital nexus were not prominent during the period in which Marx formulated his philosophy, Marx and Engels actively questioned the invasion of nature by human beings. Conceiving the class struggle as the conflicts between workers and the capitalists, or rather to believe that environmental conservation does not form part of the class struggle…such assumptions will only embolden and aid the capitalist forces.
>
> *(PMVS, Pamphlet 2008)*

The above excerpt illustrates that the leftist orientations of environmentalism are overt in the collective action frames used by movements such as the PMVS. More so, PMVS is among the many environmental mobilisations in Kerala that interpret environmentalism as a class struggle against capitalist plunder of nature. The social base of these movements among working-class members and articulation of their grievances in Marxian lexicons certainly situate these movements outside the framework of mainstream environmentalism. Additionally, the class solidarities established by PMVS among the fisherfolks and farm workers who face the fallouts from local industrial pollution vivify how the fight against environmental injustice is often a fight to secure economic livelihoods in the Global South (Satheesh, 2020c). Most importantly, the class orientations of grassroots environmental groups like the PMVS challenge the depiction of environmentalists as a middle-class collective, driven by disdain against the working class (Foster, 1993). However, despite the strong affiliation to Marxian ideology and emancipatory politics, the protest repertoires adopted by the PMVS embrace the Gandhian principles of non-violence and Satyagraha.

Not all environmental movements are NSMs

The discussion of the various characteristics of poor and working-class environmentalism in the Global South demonstrates how the existence of these movements challenges the conceptualisation of environmental movements as an NSM in the traditional sense. Emerging in the post-1960s, the NSMs were largely interpreted as a feature of post-industrial societies drawing participants markedly from the educated middle-class with postmaterial value orientations (Inglehart, 1981; Buechler, 1995; Brulle, 2000; Rootes, 2004). As explained by Rootes (2004, p. 617), environmentalism was interpreted as 'the self-interested politics of a "new class" of traffickers in culture and symbol, opposed or indifferent to the interests of those whose labour involves the manipulation of material things'. Here again, pinning down the social base of environmental movements on the educated middle-class (Cotgrove, 1982; Kriesi, 1989) negates the presence of environmental movements organised by poor and working-class people across the globe, including the environmental justice movement in the United States (US) (Mohai and Bryant, 1992; Bullard, 1993). For example, the movement for environmental justice and against environmental racism in the US, a significant political force, has often been at odds with the main environmental groups (such as Friends of Earth, The Sierra Club, etc., referred to as 'the Big Ten') (Dowie, 1996). The division between these two variants of environmentalism reflects class, race and gender. Grassroots environmental movements, such as the environmental justice movement, are led by people of colour and working-class women in contrast to the Big Ten, where the membership is dominated by white, middle-class professional men holding the centre of power (Harvey 1996). In the words of Robert Storey:

> The environmental justice movement is a social movement where race and (working) class have been enjoined through an understanding of the environment, wherein protecting the homes, health, and well-being of the urban working class and poor is the key component. As such, the environmental justice movement stands as a direct challenge to the limited conception of the environment developed by the largely White, middle-class environmental movement.
>
> *(Storey, 2004, p. 442)*

The mainstream characterisation of environmental movements as an exemplar of new politics fails to integrate diverse ecology movements into their lexicons of environmental protest. The environmental movements of the Global South, similarly, find no representation in the existing typification. As observed by Parajuli (2001, p. 96), the stark differences between 'new' movements in Western societies and a country like India are quite evident, as the former stands for 'quality of life' issues disregarding the issue of distributive justice, whereas the basic thrust of movements in India's ecological movement is to stop the monopolistic control of the rich over natural resources. Though not entirely with an economic focus, these new movements address issues related to political economy, and hence it is difficult to view them as a reflection of post-industrial and postmodern societies as suggested in the NSM literature.

Conclusion

The presence of poor and working-class environmental movements in the Global South challenges the dominant interpretations and conceptualisations that reduce environmentalism as a middle-class phenomenon. Moreover, the presence of material and environmental concerns in the movement frames of the Southern movements falsifies the monolithic claims that reduce environmentalism as driven by post-materialist value orientations. Theorising environmental movements based exclusively on western contexts and cases have rendered such framings useless to understand the rich and vibrant instances of environmental struggles in the developing world. Most importantly, the strange omission of environmental movements led by the poor and working-class people in the Global South exposes the Eurocentric and western nature of the mainstream literature that continues to build 'theories on environmentalism' based only on middle-class environmental movements. The carefully maintained silence surrounding the causes of environmental grievances, particularly within the literature on environmental, social movements, exposes how the dominant literature evades questions surrounding the role of capitalism in creating and exacerbating the environmental inequities and burdens. And this calls for the need to decentre and decolonise research on global environmental movements by bringing the struggles organised by people from the margins to the centre of academic discourse. Such an effort alone will help in creating global conceptions of environmentalism that transcend western boundaries and imaginaries.

Note

1 The post-materialist thesis by Inglehart (1990) explains environmental movements in terms of a change in cultural value towards 'quality of life' issues (Martinez-Alier 1997), according to which the poor are too poor to be green (Martinez-Alier, 1995) and that environmentalism emanates in rich countries through a shift in cultural values attributed to the post-material shift of society. This shift towards new values is explained in terms of the declining marginal utility of abundant, easily obtained material commodities.

References

Ameerudheen, T.A. (2018) '"We will save our village": In Kerala, a road project pits the CPI(M) against its supporters', *The Scroll*, 18 March. https://scroll.in/article/872272/we-will-save-our-village-in-kerala-a-road-project-pits-the-cpi-m-against-its-supporters.

Assies, W. (2003) 'David versus Goliath in Cochabamba: Water rights, neoliberalism, and the revival of social protest in Bolivia', Latin *American Perspectives*, vol 30, no 3, pp.14-36.

Bandyopadhyay, J., & Shiva, V. (1988) 'Political economy of ecology movements', *Economic and Political Weekly*, vol 23, no 24, pp. 1223-1232.

Baviskar, A. (1995) *In the Belly of the River: Tribal Conflicts over Development in the Narmada Valley*. Oxford University Press, Oxford.

Baviskar, A. (2005). 'Red in tooth and claw? Looking for class in struggles over nature, in Ray, R. and Katzenstein, M.F. (ed) *Social Movements in India: poverty, power and politics*, Rowman & Littlefield, New York.

Bijoy, C.R. (2006) 'Kerala's Plachimada struggle: A narrative on water and governance rights', *Economic and Political Weekly*, vol 41, no 41, pp. 4332–4339.

Brulle, R.J. (2000) *Agency, Democracy, and Nature: The US Environmental Movement from a Critical Theory Perspective*. MIT Press, Cambridge, MA.

Buechler, S.M. (1995) 'New social movement theories', *The Sociological Quarterly*, vol 36, no 3, pp. 441–464.

Bullard, R.D. (ed.) (1993) *Confronting Environmental Racism: Voices from the Grassroots*. South End Press, Boston, MA.

Chandran, R. (2020) 'Deadly land conflicts seen rising as threat from industry grows', Reuters, 29 July. https://www.reuters.com/article/us-global-landrights-violence-idUSKCN24U005

Cotgrove, S. (1982) *Catastrophe or Cornucopia*. Wiley, New York.

Cowie, S. (2019) 'Brazilian forest guardian killed in illegal ambush', *The Guardian*, 2 November.

Dowie, M. (1996) *Losing Ground: American Environmentalism at the Close of the Twentieth Century*. MIT Press, Cambridge, MA.

Dwivedi, R. (2001) 'Environmental movements in the global south: Issues of livelihood and beyond', *International Sociology*, vol 16, no 1, pp. 11–31.

Erdős, L. (2019) 'The sacrifice of Berta Cáceres', in Erdős, L. (ed) *Green Heroes*. Springer, Cham, pp. 199–202. https://doi.org/10.1007/978-3-030-31806-2_39

Foster, J.B. (1993) 'The limits of environmentalism without class: Lessons from the ancient forest struggle of the Pacific Northwest', *Capitalism Nature Socialism*, vol 4, no 1, pp. 11–41.

Guha, R. and Gadgil, M. (1989) 'State forestry and social conflict in British India', *Past and Present*, vol 123, pp. 141–177.

Gadgil, M., & Guha, R. (1993) *This Fissured Land: An Ecological History of India*, University of California Press, Berkeley.

Gadgil, M. and Guha, R. (1994) 'Ecological conflicts and the environmental movement in India', *Development and Change*, vol 25, no 1, pp. 101–136.

Global Witness, (2020) 'Defending Tomorrow: The climate crisis and threats against land and environmental defenders', Report published by Global Witness, July 2020. www.globalwitness.org

Greenfield, P. and Watts, J. (2020) 'Record 212 land and environmental activists killed last year', *The Guardian*, 29 June 29.

Grove, R. (2002) 'Climatic fears: Colonialism and the history of environmentalism', *Harvard International Review*, vol 23, no 4, p. 50.

Guha, R. (1985) 'Scientific forestry and social change in Uttarakhand', *Economic and Political Weekly*, vol 20, no 45/47, pp. 1939–1952.

Guha, R. (2000) *The Unquiet Woods: Ecological Change and Peasant Resistance in the Himalaya*. University of California Press, Berkeley, CA.

Guha, R. (2002) 'Environmentalist of the poor', *Economic and Political Weekly*, pp. 204–207.

Harvey, D. (1996). *Justice, Nature and the Geography of Difference*. Blackwell, Oxford.

Heller, P., Harilal, K.N. and Chaudhuri, S. (2007) 'Building local democracy: Evaluating the impact of decentralization in Kerala, India', *World Development*, vol 35, no 4, pp. 626–648.

Inglehart, R. (1981). 'Post-materialism in an environment of insecurity', *The American Political Science Review*, vol 75, no 4, pp. 880–900.

Inglehart, R. (1990) *Culture Shift in Advanced Industrial Society*, Princeton University Press, Princeton, NJ.

Isaac, T.T. and Harilal, K.N. (1997) 'Planning for empowerment: People's campaign for decentralised planning in Kerala', *Economic and Political Weekly*, vol 32, no 1/2, pp. 53–58.

Karan, P.P. (1994) 'Environmental movements in India', *Geographical Review*, vol 84, no 1, pp. 32–41.

Kriesi, H. (1989) 'New social movements and the new class in the Netherlands', *American Journal of Sociology*, vol 94, no 5, pp. 1078–1116.

Kurien, J. (1991) *Ruining the Commons and Responses of the Commoners: Coastal Overfishing and Fishermen's Actions in Kerala State, India*. United Nations Research Institute for Social Development, Geneva.

Lakhani, N. (2020) *Who Killed Berta Caceres?: Dams, Death Squads, and an Indigenous Defender's Battle for the Planet*. Verso, New York.

Lin, S.G. (1992) 'Shankar Guha Niyogi: Beyond conventional trade unionism in India', *Bulletin of Concerned Asian Scholars*, vol 24, no 3, pp. 16–25.

Martinez-Alier, J. (1995) 'The environment as a luxury good or "too poor to be green?"', *Ecological Economics*, vol 13, no 1, pp. 1–10

Martínez-Alier, J. (1997) 'Environmental justice (local and global)', *Capitalism Nature Socialism*, vol 8, no 1, pp. 91–107.

Martinez-Alier, J. (2003) *The Environmentalism of the Poor: A Study of Ecological Conflicts and Valuation*. Edward Elgar Publishing, Northampton, MA, USA.

Méndez, M.J. (2018) '"The river told me": Rethinking intersectionality from the world of Berta Cáceres', *Capitalism Nature Socialism*, vol 29, no 1, pp. 7–24.

Mohai, P., and Bryant, B. (1992) 'Environmental injustice: weighing race and class as factors in the distribution of environmental hazards', *The University of Colorado Law Review*, vol 63, pp. 921.

Montrie, C. (2011) *A People's History of Environmentalism in the United States*. A&C Black, New York.

Montrie, C. (2018). *The Myth of Silent Spring: Rethinking the Origins of American Environmentalism*. University of California Press, Oakland, CA.

Na'Allah, A.R. (ed.) (1998) *Ogoni's Agonies: Ken Saro-Wiwa and the Crisis in Nigeria*. Africa World Press, Trenton, NJ.

Nilsen, A.G. (2008) 'Political economy, social movements and state power: A Marxian perspective on two decades of resistance to the Narmada dam projects', *Journal of Historical Sociology*, vol 21, no 2–3, pp. 303–330.

Nilsen, A.G. (2010) *Dispossession and Resistance in India: The River and the Rage*. Routledge, New York.

Obi, C.I. (1997) 'Globalisation and local resistance: The case of the Ogoni versus Shell', *New Political Economy*, vol 2, no 1, pp. 137–148.

Olivera, O. and Lewis, T. (2004) *Cochabamba!: Water War in Bolivia*. South End Press, Cambridge, MA.

Osaghae, E.E. (1995) 'The Ogoni uprising: Oil politics, minority agitation and the future of the Nigerian state', *African Affairs*, vol 94, no 376, pp. 325–344.

Parajuli, P. (2001). 'Power and knowledge in development discourse: New social movements and the state in India', in Jayal, N.G. (ed) *Democracy in India*, Oxford University Press, Oxford, pp. 258-88

Parayil, G. (1996) 'The "Kerala model" of development: Development and sustainability in the Third World', *Third World Quarterly*, vol 17, no 5, pp. 941–958.

Raman, K.R. (2005) 'Corporate violence, legal nuances and political ecology: Cola war in Plachimada', *Economic and Political Weekly*, vol 40, no 25, pp. 2481–2486.

Rangan, H. (2000) *Of Myths and Movements: Rewriting Chipko into Himalayan History*. Verso, New York.

Revkin, E. (2004) *The Burning Season: The Murder of Chico Mendes and the Fight for the Amazon Rain Forest*. Island Press, Washington.

Rootes, C. (2004) 'Environmental movements', in D.A. Snow, S.A. Soule and H. Kriesi (eds.), *The Blackwell Companion to Social Movements*. Blackwell Publishing, Bridgewater, NJ.

Satheesh, S. (2020a) 'Moving beyond class: A critical review of labor-environmental conflicts from the global south', *Sociology Compass*, e12797.

Satheesh, S. (2020b) 'The pandemic does not stop the pollution in River Periyar', *Interface: A Journal on Social Movements*, vol 12, no 1, pp. 250–257.

Satheesh, S. (2020c) 'Red-green rows: Exploring the conflict between labour and environmental movements in Kerala, India', PhD Thesis, University of South Florida, Tampa.

Seetharaman, G. (2018) 'The story of one of the biggest land conflicts: No mine now, but is it all fine in Niyamgiri?', *The Economic Times*, 18 April.

Shiva, V. (1991) *The Violence of the Green Revolution: Third World Agriculture, Ecology and Politics*, Zed Books, New York.

Shiva, V. and Bandyopadhyay, J. (1986) 'The evolution, structure, and impact of the Chipko movement', *Mountain Research and Development*, vol 6, no 2, pp. 133–142.

Shrivastava, A. and Kothari, A. (2012) *Churning the Earth: The Making of Modern India*. Penguin Viking, New Delhi.

Smythies, E. A. (1925) *India's Forest Wealth*, Humphrey Milford, London.

Storey, R. (2004) 'From the environment to the workplace and back again? Occupational health and safety activism in Ontario, 1970s–2000', *Canadian Review of Sociology/Revue Canadienne de sociologie*, vol 41, no 4, pp. 419–447.

Taylor, D.E. (1993) 'American environmentalism: the role of race, class and gender in shaping activism 1820–1995', *Race, Gender & Class*, vol 5, no 1, pp. 16–62.

Temper, L., Daniela, B. and Martinez-Alier, J. (2015) 'Mapping the frontiers and front lines of global environmental justice: The EJAtlas', *Journal of Political Ecology*, vol 22, no 1, pp. 255–278.

The New Indian Express (2017). 'Kandankali protest pose another challenge to CPIM', 22 November.

Watts, J. and Vidal, J. (2017) 'Environmental defenders being killed in record numbers globally, new research reveals', *Chain Reaction*, vol 130, p. 40.

5

WORKING-CLASS PEOPLE, EXTINCTION REBELLION AND THE ENVIRONMENTAL MOVEMENTS OF THE GLOBAL NORTH

Karen Bell

Introduction

For decades, mainstream environmental organisations of the Global North have been criticised for failing to engage Black, Asian and Minority Ethnic (BAME), low-income and working-class communities (e.g. Taylor, 1997; Agyeman, 2001; Burningham and Thrush, 2001; Capacity Global, 2009; Taylor, 2016; Bell, 2020). Numerous academics and commentators have highlighted the primarily middle-class composition of a range of mainstream environmental organisations, including, for example, Transition Towns (e.g. Grossmann and Creamer, 2017), Greenpeace (e.g. Harter, 2004) and Earth First! (e.g. Loomis, 2016). Recently, Extinction Rebellion (XR) has become a particular target for criticism regarding its lack of diversity and inclusiveness (e.g. Hinsliff, 2019; Shand-Baptiste, 2019; Josette, 2019). In particular, the United Kingdom (UK) media have fixated on a critique of XR as being primarily middle-class, White and not representing the views of society as a whole. For example, it was described as a 'loopy middle-class doomsday cult' by Sky presenter, Carole Malone (Malone, 2019) and Spiked's Brendan O'Neill depicted the group as 'an anti-working-class movement' (O'Neil, 2020). While these UK media criticisms may be somewhat extreme, academic research tends to confirm the demographics of the group in terms of class, for example, with fewer than 10% of XR activists identifying themselves as working class in a recent survey (Saunders et al., 2020). In this chapter, I explore the reasons for this lack of diversity and inclusion in the mainstream environmentalism of the Global North. I then identify some steps to address it, drawing on my personal experience as a life-long environmental activist from a working-class background; my work experience as a community development worker in disadvantaged communities; and relevant academic research, including my own.

After first defining 'class', the chapter goes on to critique the myth that working-class people are unable or unwilling to take action on the environment. This follows with a discussion of the barriers to engaging with mainstream environmentalism for working-class people, focussing on the problematic tactics and discourse of XR as a recent example of the ongoing existence of such barriers. Finally, as suggestions for alternative forms of organising for sustainability, I discuss two forms of 'working-class environmentalism', that is, the trade union health and safety movement and the environmental justice movement.

What is class?

To add clarity to the discussion, it is important to articulate what is meant here by 'class'. Like many other equalities concepts, 'class' is a contentious notion and, as a result, it has been defined in numerous different ways. It has, for example, been considered as a division based on exploitation, income, education, status, occupation, cultural signals, subjective affiliation, lifestyles or some combination of these (for debates, see Wright, 2001; Goldthorpe and Chan, 2007; Bradley, 2014). There are particularly divisions between those who adopt a 'relational' view of class and those who take a 'gradational' position. Marxism is a typical relational view, whereby class is seen in terms of our 'relationship to the means of production'. From this perspective, classes are opposed to each other in a hierarchy of power, privilege and financial resources. By contrast, a 'gradational' approach considers class in terms of levels of education, occupation, income and status. As a gradational approach, the UK National Statistics Socio-economic Classification (NS-SeC), for example, distinguishes eight classifications based on occupation, from higher managerial and professional occupations to never worked or long-term unemployed. More recent sociological interpretations have focussed on 'cultural class analysis' (Atkinson, 2010; 2015), emphasising the differing conditions of life arising from distinctive class cultures and the consequent power dynamics (e.g. Skeggs, 2004; McKenzie, 2015). The inter-generational aspect of class is also important (CLASS, 2017) with static or declining intergenerational occupational mobility now apparent in many countries (Social Mobility Commission, 2017; OECD, 2018). For example, a recent report on the outcome of social mobility policy in the UK over the last 20 years showed that our class position is, even now, usually determined by that of our parents (Social Mobility Commission, 2017). For the purposes of this chapter, class will be conceived of as a synthesis of the gradational and cultural definitions, that is, in terms of current and inherited wealth, income, occupation, status, 'recognition' and valuing.

In the UK, social class is often left out of discussions about diversity and inclusion, in part, because it is not part of equalities legislation here. There is no legal requirement that working-class people should not be treated less favourably than their middle-class counterparts. However, we know that class continues to be of great relevance for determining life outcomes, including with regard to education (Reay, 2017), occupation (Toft and Friedman, 2020), income (Laurison and Friedman, 2016) and physical and mental health (Eisenberg-Guyot and Prins, 2019; Muntaner et al., 2015). In relation to environmentalism, as this chapter outlines, class can determine the quality of our environment, our chances of having a say in environmental decision-making and our likelihood of being included in mainstream environmentalism. Therefore, class is enormously relevant to considerations of diversity and inclusion in environmentalism.

Do working-class people care about the environment?

The lack of working-class people in mainstream environmental movements has often been explained away in terms of this group's lack of concern about the environment. For some time, it was considered that environmentalism was only the preoccupation of middle-class citizens in wealthy, highly industrialised nations of the Global North. In particular, Inglehart's (1977) 'post-materialist values theory' argued that, with increased affluence, concern for quality-of-life issues, such as free speech, liberty, and environmental protection (post-materialist values), increases. This, he argued, arises only after individuals have met their more basic materialist needs for food, shelter and safety. Similarly, the 'affluence hypothesis' (Franzen, 2003) assumes a direct link between affluence and environmental concern. This theory was supported by a number of studies that showed a positive association between higher socio-economic status and environmental concern (e.g. Gelissen, 2007; Marquart-Pyatt, 2008; Franzen and Meyer, 2010; Nawrotzki and Pampel, 2012; Franzen and Vogl, 2013).

However, these ideas have been challenged by other studies that found that affluence is not consistently positively correlated with environmental concern (e.g. Dunlap and Mertig, 1997; Dunlap and York, 2008) and that working-class people are at least as concerned about the environment as their wealthier counterparts (e.g. Uyeki and Holland, 2000; Power and Elster, 2005). While focussing mainly on the Global South, Joan Martinez-Alier (2003) and Sunita Narain (2013), among others, describe local resistance to environmental destruction in low-income and indigenous communities as the 'environmentalism of the poor'. In these cases, they observe 'the poor' as being highly motivated to defend the environment because they are strongly aware that it supports their livelihoods, well-being and survival. Even in the high consumption countries of the Global North, poor and working-class people tend to have a less damaging impact on the environment because their, generally lower and more insecure, incomes tend to mean they consume much less than their wealthier counterparts. Hence, the overall carbon footprints for people on low incomes tend to be lower (see, e.g., Pang et al., 2019).

The environmental justice literature also supports the idea that low-income communities are often very active environmental campaigners and custodians (Martinez-Alier, 2003; Pellow, 2007; Pellow, 2018). For example, the global Environmental Justice Atlas has found that mobilisations for more sustainable and socially just uses of the environment are apparent across all income groups around the world (Scheidel et al., 2020).

One of the main flaws in the studies claiming that low-income groups are not concerned about the environment is that they do not take into account different ways of expressing environmental concern. These studies focus primarily on green consumerism and climate activism. However, the generally lower and more insecure incomes of working-class people restrict their ability to carry out green activities which have a direct or indirect financial cost, for example, buying organic food, purchasing longer lasting 'quality' products, eco-tourism holidays and taking time off from work to engage in climate protests (Bell, 2020). The idea that environmentalism is predominantly a middle-class phenomenon draws on a very narrow definition of 'environment', focussed principally on climate change, preservation of wilderness and biodiversity (Bullard and Wright, 1992; Di Chiro, 1996; Allen et al., 2007). However, working-class people and the global poor have tended to focus on maintaining environments that are adequate for immediate physical survival (Bell, 2020; Satheesh, 2020). Pulido (1998, p. 30) calls this an 'environmentalism of everyday life'.

Many studies indicate that working-class concerns may be different from those of middle-class people (e.g. Burningham and Thrush, 2001, 2003; Agyeman, 2002; Bell, 2020). For example, Burningham and Thrush (2003), investigating how people living in disadvantaged communities talk about and experience environmental inequality, noted a focus on everyday environmental concerns and a lack of 'the language of environmentalism' in their narratives. Instead, the residents focused on health and safety at home and in the streets around them, as well as the social problems that impacted their lives (Burningham and Thrush, 2003). These communities were both interested in and active on environmental issues. Working-class environmental concerns and activism need to be recognised as forms of 'environmentalism' if the movement is to be inclusive.

Analysts have noted that environmental concerns are strongly influenced by the surrounding environmental conditions (e.g. Abramson and Inglehart, 1995; Rohrschneider, 1990). Working-class, low-income and other disadvantaged groups often have a direct experience of environmental degradation due to their greater proximity to environmental harms. For example, in the UK, low-income groups and working-class people experience the worst air quality (Mitchell and Dorling, 2003; Walker et al., 2003; Pye el al., 2006; Milojevic et al., 2017), greater toxicity from proximity to waste sites (Wheeler, 2004; Fairburn et al., 2009; Richardson et al., 2010), less access to green space (Defra, 2011; Burt el al., 2013), food poverty (Taylor and Loopstra, 2016), fuel poverty (IPPR, 2018), inadequate transport infrastructure (Lucas, 2012) and are more likely to be impacted by all the risks associated with global climate change, including flooding (Walker et al., 2007; Walker and Burningham, 2011; Oxfam, 2014). All of these additional risks and threats directly undermine the health and lifespan of these communities. Hence, working-class people become sensitised to environmental issues, even if they sometimes feel powerless to do anything about it. Working-class people have also looked to more collective solutions to these problems, such as regulation, rather than individual responses, such as lifestyle change (see Bell, 2014; Bell, 2020).

Overall, then working-class people do care about the environment, though their ways of considering and addressing the associated problems can go unrecognised as 'environmentalism'. This provokes the questions that, if working-class people are environmentalists, why are they so unlikely to join mainstream environmental organisations? My recent research and personal experience suggest that the working-class absence from mainstream environmental organisations is a direct result of the lack of inclusive practices on the part of these organisations. The next section explores these practices and how to change them.

How can mainstream environmental organisations become more inclusive to working-class people?

First, it is important to emphasise that environmental groups come in many shapes and forms, and therefore some will be more difficult for working-class people to be a part of than others. Also, it is very evident that some key mainstream environmental organisations now recognise the need to be more inclusive of working-class people and are working on addressing it. I have been contacted by a number of these organisations over the last year for advice on how to be more inclusive. Moreover, some of the leaders of these organisations have been committed to change on this issue for some time. For example, Craig Bennett, Chief Executive Officer (CEO) of Friends of the Earth (FoE) since 2015, spoke of this aspiration when taking on his leadership role with FoE, commenting:

> I feel passionately that, to up our game, the environmental movement needs to make sure it is not just a white middle-class movement. We need to make a really big effort to look at how we can reach out to different communities and create opportunities for them to campaign on the issues that matter to them.
>
> *(Bennett, 2015)*

Despite this goodwill, many of these organisations are struggling to find a way forward. Therefore, the next section offers some basic advice. Alongside the academic literature and personal experience, it draws on some of my recent research (Bell and Bevan, 2021, forthcoming), investigating class issues in relation to XR. The latter involved 40 in-depth semi-structured interviews with working-class people across England and Wales. Triangulated with the other sources, it helps to illustrate more general points.

Be proactive in finding opportunities for mutual solidarity

The first step to becoming inclusive as an environmental organisation is having the intention to make changes in this direction, being prepared to undertake new activities and going beyond your comfort zone. Grossmann and Creamer's (2017) study of the Tooting Transition Town group describes how, although the group had an 'open-door' policy, it did not actively set out to recruit a diverse range of participants, follow through on their ideas, or consider their particular needs. An open-door policy is inadequate as it does not take into account the barriers to participation; therefore, as the researchers noted, this 'passively inclusive' approach failed to achieve diversity. My recent and prior research indicates that one of the main reasons as to why working-class people do not get involved with environmental groups is that they do not know they exist or know little about their achievements and successes (Bell, 2020; Bell and Bevan, 2021, forthcoming). Working-class people are expected to find the environmental organisations them-selves, make whatever sacrifices are necessary to get involved with them and then 'fit in'. Therefore, it would be useful for mainstream environmental organisations to (1) find out about existing working-class networks and activism; (2) go into those working-class spaces – not as a recruiter or teacher, but to learn about the local environmental and social issues and work with people to develop constructive relationships over time; (3) find common causes and campaign on them together bringing a mindset of 'mutual solidarity', rather than 'helping'.

Consider the impact of your discourse and tactics on working-class people

Another important step would be for the mainstream environmental organisations to examine their discourses and tactics. A key action would be to ask working-class people what they think of these via their representative and accountable organisations, such as trade unions and community groups. In our recent research on class and race in relation to XR (Bell and Bevan, 2021, forthcoming), the working-class interviewees, especially those of BAME backgrounds, felt very uncomfortable and even angry about XR's tactics because they tend to involve illegal and disrup-tive activities. They, particularly, criticised the actions blocking the trains in London since they prevented people getting to work and targeted a relatively low carbon transport method. For example, Jake remarked:

> Their last protest in London – I was actually working in London at the time. The chaos it caused me travelling … I had to leave at five o'clock in the morning and I wasn't getting home until 12 o'clock at night because of the blockages they were causing. They were causing delays everywhere. To be honest, I didn't feel safe leaving the venue I was at until the crowds had gone.
> *(Jake, interview, 27 August 2020)*

A key critique of XR, especially by the BAME working-class interviewees, was in relation to their concerns about engaging in criminal activity and interacting with the police. It is important to use tactics that working-class people can support. The strategy of mass arrest is very difficult for working-class people to enthusiastically embrace. Although XR leaders have said this is optional, they still often portray it as admirable and signifying depth of commitment. As Wretched of the Earth (2019) has so clearly articulated, this approach fails to take into account the reality for BAME people. For people from BAME and working-class backgrounds, there is no margin for error. Although there must be a place for disobedience in any campaign for social change, the way some of the activists spoke about this disruptive activity seemed to imply that anyone could do the same. For example, when challenged on Radio 4's Today programme about the disruption to people's lives that XR was causing, Gail Bradbrook, one of the organisers, responded: 'they should take some time off work and come and join us' (BBC Radio 4, 18 April 2019).

It is important to think about the targets and victims of campaigning actions. Harter (2004), for example, notes that the middle-class founding members and subsequent leaders of Greenpeace tended to target not only the responsible companies in their hunting and logging campaigns, but also working-class communities struggling for survival. He stated that '[d]espite their claims to be democratic, universal, and above class interests, the methods they choose often have very real negative effects on one particular group: the working-class' (Harter 2004, p. 92). In the late 1990s, for instance, instead of developing a campaign strategy with logging communities, Greenpeace blockaded the loggers, preventing them from working and earning a living (Harter, 2004). XR protesters replicated this mistake when they prevented Smithfield Market traders from operating by turning the site into a vegan pop-up (Morris, 2019). This prevented the market workers earning a living that day and led to substantial food waste.

When there are no immediate alternative options for employment, criticising these jobs without discussing alternative livelihoods with the workers is very likely to alienate them. It is not only financial concerns that can make working-class people unsupportive of environmental campaigns, though, there is also the important aspect of pride and dignity resulting from occupation. Joe Uehlein of the US Labor Network for Sustainability asserts that environmentalists 'have never been able to understand the primacy of work in people's lives so their messaging is tone deaf to the needs and aspirations of working people' (Labor4sustainability, 2017). Räthzel and Uzzell (2012, p. 84) highlighted that work provides dignity, identity and solidarity so that when industries are attacked as environmentally harmful, those who work in those industries will also feel attacked.

Criticise the powerful, not those without options

Some environmentalists focus on individual solutions as the answer to environmental degradation, such as encouraging people to choose to engage in 'green consumption' (e.g. organic food or eco travel), or 'voluntary simplicity', that is, to reduce consumption in general and live more simply. These are supported by a whole range of lifestyle and grassroots groups and campaigns, including Slow Food, Permaculture, eco-villages, Transition Towns, community gardens, eco-building, farmers' markets, local currencies, co-housing projects and skill-sharing networks. However, many of these fail to engage with working-class communities and are inaccessible in terms of time, finances, skills, contacts and everyday stresses (Bell, 2020).

When mainstream environmentalists primarily emphasise the need to modify our individual behaviour and ways of earning a living, they can come across as un-empathic and uncaring. While it is indisputable that we need to consume less resources and emit less waste as a species to avoid irreversibly overstep planetary boundaries (Steffen et al., 2015; IPCC, 2018), putting so much emphasis on individual choices ignores the limited options that are available to many working-class people. In general, environmentalism is often attributed to individual dispositional factors, such as attitudes and values, rather than contextual and structural factors, such as income, time and availability of green options (Uzzell and Räthzel, 2009). Hence, in the policy domain, interventions to encourage people in low-income areas to increase participation in recycling have often been based on providing information (assuming lack of knowledge), rather than making recycling more convenient (removing situational barriers) (Jackson, 2005). Environmental concern, although complex (Knight et al., 2012), is strongly influenced by structural factors and cultural contexts (Uzzell and Räthzel, 2009).

The working-class people I have interviewed in my research are keenly aware that, lacking options, they are unable to make some green behaviour changes. They also feel that greens show no empathy or understanding of their limited situation (Bell, 2020: Bell and Bevan, 2021, forthcoming). Hence, lifestyle approaches can, and often do, alienate working-class people by assuming that everyone has the same options to carry out environmentally virtuous acts. There is, generally, a common assumption among middle-class people that their experiences are universal. Like many powerful groups, they often seem to be unaware of their privilege. One aspect of this is the question of financial cost – often a major barrier to working-class inclusion.

Do not be classist

Where working-class people do engage with environmental organisations, they face micro-aggressions and classism. I have personally seen this in the many environmental meetings I have attended over the years. For example, there is a pervasive right to correct that middle-class people appear to hold. In most of my research interviews, working-class people reported at least one experience of being 'told off' by middle-class people in the environmental groups they had engaged with. In general, working-class people say that environmentalists tend to be sanctimonious and often treat them in a patronising way (Bell, 2020). Similarly, in our recent research about XR, almost all our research participants said they would feel uncomfortable around XR activists as they did not feel the group could relate to them or their communities. For example, Dave, a security guard, said:

> I see them as a, as a largely middle-class group of people… If you try to take some of those people on to the council estates in Bristol, or even Liverpool or anywhere else, or outside of their little bubble, I think, they wouldn't be able to relate for five minutes… Or they'd probably just fall back on their prejudices….
> *(Dave, interview, 2 August 2020)*

Many of the working-class research participants felt that middle-class environmentalists criticised them for their lifestyles, jobs and culture. They said that middle-class people tend to look upon the working-class as needing to be helped, rather than seeing them as equal or having their own qualities and ways of contributing. This also resonates with my own experience with mainstream environmental groups and programmes. For example, in one mainstream environmental organisation, I was nominated as the 'Diversity Champion' for the local branch. In this role, I was responsible for helping to ensure equalities, diversity and inclusion in their campaign activities. However, I found it impossible to get the group to make the changes that would achieve these gaols. For example, when I suggested that we should undertake activities on the council estates on the outskirts of the city (where I, myself, live), I was told that these places were 'too difficult to get to'. When I proposed that we should set up a system to monitor diversity in the local group and national organisation, the group decided it would not be a good idea because 'it might upset someone'. On suggesting that we should campaign to change local government policy on waste collection because many working-class people could not afford to pay for the chargeable services, they said those working-class people 'need educating'. I continued to raise these, and other, issues at my local branch but all were dismissed. One middle-class woman said that society is too diverse now for class to have any meaning. Several people said that class was not an important or relevant issue. These occurrences are consistent with the literature that describes how those from privileged groups tend to deny the existence of oppression (see Goodman, 2011). Problems raised by members of the oppressed group are routinely dismissed and the person raising the issues is labelled as 'over-sensitive' or a 'troublemaker'. This whole experience was so frustrating for me that I eventually decided to leave the group.

Consistent with studies that note a 'pathologisation of working class people' (e.g. see Walkerdine, 2020, on the Brexit vote), the working-class people I have spoken to in my research and practice consider that many aspects of their lives and culture are negatively judged by environmentalists. In our recent research on XR, a number of interviewees expressed anxiety that they would be criticised for their lifestyle and choices, as Gabrielle comments here:

> To me, I get the impression that they are a lot of students, uni type students. That's the impression I was getting anyway... They're like vegetarian type people, vegan type people. There's nobody there like people who might eat meat. [laughs] ....
>
> *(Gabrielle, interview, 15 August 2020)*

Such comments illustrate the anxieties that working-class people have about being rejected and demeaned by middle-class environmental organisations. This is a major barrier to overcome if these organisations are going to recruit or retain working-class people.

To a large extent, classism in the environmental movement is similar to classism in society. Therefore, it would be useful for middle-class environmentalists to read the work of working-class academics that specialise in this area (e.g. Lisa McKenzie, Diane Reay, Valerie Walkerdine and Bev Skeggs).

Focus on environmentalism as a chance to make everyone's life better

In general, reductions in income, work and consumption can appear threatening to working-class communities who often struggle to hold on to their jobs and material standard of living. A great deal of environmental discourse is focused on consuming less, cutting-back and self-denial. Part of this has been a tendency for middle-class environmentalism to fetishise poverty or consider it a game or adventure. But, as Vosper (2016) states, this 'is an insult to folk who have no choice ... unlike middle-class people we don't have a safety net... Flirting with poverty as a lifestyle choice is not the same as growing up in poverty' (Vosper, 2016, n.p.). Working-class people tend to want to hide their poverty and present themselves as 'doing OK'. McGarvey (2017, p. 23), reflecting on his working-class childhood, notes 'the deep sense of shame many of us felt about our poverty' and how buying all the goods to appear better off was expensive 'but the price of looking poor was higher'. It is the inequalities in society that drive this consumerism (Wilkinson and Pickett, 2010, 2019) as everyone becomes desperate to demonstrate adequate status to avoid rejection (Veblen 1994 [1899]). To prevent this, sustainability policies need to reduce economic inequality, as discussed in the final chapter of this book.

An orderly economic contraction or degrowth may be necessary but at the global level it has to be discussed and achieved in a way that improves working-class lives. Degrowth or post-growth can benefit working-class people through linking it to a much wider structural change that will bring about greater equality. In the current economic and political structure, such an economic contraction would, as we have seen from the coronavirus (COVID-19) lockdowns, almost certainly lead to companies going out of business and increasing unemployment levels. These kinds of discourses need to be thought through so that they do not harm working-class people or alienate them from environmentalism. A positive attempt to do this is a recent work by Mastini et al. (2021), which has looked at synthesising Green New Deal and degrowth approaches into a 'Green New Deal without growth'. This would include public investments for financing the energy transition, industrial policies for the decarbonisation of the economy, socialising the energy sector, and expanding the welfare state to increase social protection.

Recognise and support working-class environmentalism

As discussed earlier, middle-class environmentalists and their organisations have not recognised the kinds of environmentalism that working-class people do engage in and have done since long before their own organisations and programmes came into existence. Recognising these working-class achievements will enable environmental organisations to build on them. In many respects, rather than try to entice working-class people into middle-class organisations, it might be more helpful for middle-class environmental activists to consider using their resources to support working-class movements and organisations that work on environmental issues. The next section discusses two of these: the trade union health and safety movement and the environmental justice movement.

Working-class environmentalism

Trade union health and safety

The Trade Union Health and Safety movement has brought awareness of the environmental contaminants present in workplaces and communities since before mainstream environmentalism existed. These working-class trade unionists have commissioned research and put pressure on companies and governments to adopt and enforce strict environmental policies and regulations through collective organisation, barefoot epidemiology and self-education (Bell, 2020; Lerner, 2012). Environmental standards are often developed only when illnesses among workers become too glaringly obvious to ignore. There are many historical examples of the achievements of trade union health and safety activism, including numerous campaigns to reduce the environmental determinants of poor health in the workplace and community. Studies on byssinosis (e.g. Bowden and Tweedale, 2003), silicosis (e.g. Bufton and Melling, 2005), asbestosis and other conditions show how concern for environmental health at work has had an impact on wider society and ecology. For example, in 1899 the UK Women's Trade Union League (WTUL) led a campaign against the use of lead in the manufacture of pottery, which was causing blindness, convulsions and death among the mostly female workforce, as well as affecting their unborn children (WTUL, 1899). As a result of their work, we no longer use lead-glazed plates and cups.

These struggles for health and safety continued throughout the 20th century to the present day. For example, in the 1960s and 1970s in the United States, oil, chemical, atomic, steel and farm workers unions campaigned against environmental risks. This activism led to significant new environmental regulations in the United States, such as the 1970 Clean Air Act and the 1972 Clean Water Act (Gottlieb, 1993). Similarly Macphee (2014) describes how, in the 1960s, workers and residents in Yellowknife, Canada, suspected that the arsenic released as a by-product of local gold mining was responsible for the increase in local cancer rates. The workers in Yellowknife, allied with federal and regional First Nations organisations, conducted their own joint study of their members to assess the health implications of arsenic exposure. Some of the samples they collected contained arsenic levels 50 times above the World Health Organization's designated 'safe' standard. As a result of their campaigns to expose these issues, in 1978, the Canadian government eventually announced its intention to create stricter regulations surrounding arsenic emissions from gold mining at a national level.

Trade union environmentalism has often gone beyond protecting workers in the union and local community to wider action on global issues, such as climate change. For example, FILLEA-CGIL, the biggest Italian construction union, has moved beyond a narrow focus on immediate growth in production, currently calling for the end of the construction of unnecessary new buildings. In a statement they explained:

> At stake, is not just the future of work in our sector, but also the future of the countryside…we have to stop building for building's sake…we need a new urban strategy, capable of drastically reducing the consumption of land and the use of cement.
>
> *(FILLEA-CGIL in Clarke and Sahin-Dikmen, 2018, p. 1)*

They called for the reuse and restoration of existing buildings and for legislation to prevent the extent of speculation and using housing as an investment (in Clarke and Sahin-Dikmen, 2018). Calls for 'Just Transition' are now emblematic of these wider trade union aspirations, through which the labour movement has played a key role in developing equitable policies on climate change and transitioning to sustainability, more generally (see, e.g., Stevis et al., 2020).

The environmental justice movement

Environmental justice is fundamentally about achieving a healthy environment for all social groups, now, and in the future (i.e. substantive environmental justice); an equitable distribution of environmental 'goods' and protection from environmental harms for all socio-economic groups (i.e. distributive environmental justice); and fair, participatory and inclusive structures and processes of environmental decision-making (i.e. procedural environmental justice) (Bell, 2014). The term 'environmental justice' was first recorded as being used in the 1960s when, predominantly working-class, African American workers in Detroit protested against pollution in and around the car factories (Rector, 2014). More often, the origins of the term are located in the 1980s when 'People of Color' in the United States began to protest about hazardous and polluting industries being disproportionately located in their neighbourhoods. The particular 'spark' is usually considered to be the proposed siting of a toxic polychlorinated biphenyl (PCB) -contaminated soil landfill in Warren County, North Carolina. Following a lack of response to the concerns they expressed through the usual democratic channels, local residents finally decided to block the road to prevent the contaminated PCB soil being deposited (Bullard, 1990). Numerous studies began subsequently to provide statistical evidence of a significant correlation between the location of hazardous waste sites and the proximity of low-income and/or BAME communities (e.g. Bullard, 1983; US GAO, 1983; UCC CRJ, 1987). Bullard (1983), for example, found that, in Houston, six of the eight city-owned waste incinerators and three of the four landfills were placed in predominantly Black neighbourhoods. A national network of environmental justice activists went on to organise the First National People of Color Environmental Leadership Summit in Washington, DC in 1991. The resulting Summit document, 'The Principles of Environmental Justice', included the right to a safe and healthy work environment and the cessation of the production of all toxins, hazardous wastes, and radioactive materials (First National People of Color, 1991).

Though the Bush and Trump administrations have undermined the US environmental justice movement to an extent, at its height, this movement was able to prevent the siting of numerous hazardous facilities in working-class communities (see, e.g., Pellow, 2007). As a result of these mobilisations, 'environmental justice' struggles have now been taken up by social movements, and, in some cases, policymakers around the world (Bell, 2014; Walker, 2012).

In the UK, environmental justice, as a concept, emerged in the mid-1990s, predominantly via the academic community. Policymakers have also intermittently focused on it, mostly in terms of 'environmental inequality' (Bell, 2008). However, working-class communities in the UK rarely see themselves as environmental justice campaigners or even environmentalists, even though they continually mobilise to defend their communities against environmental hazards (Bell, 2020). There are numerous accounts of environmental and social justice actions in working-class communities (e.g. Smith, 1997a, 1997b; Harley and Scandrett, 2019). The environment is a key issue for communities since it relates to health, well-being, jobs and relationships. As a former community development worker, I regularly supported working-class people to address local socio-environmental challenges. This included making local derelict land safe and accessible for play and leisure; organising youth camps for inner city young people to visit the countryside (Albany Youth and Community Centre, St Pauls); setting up a food coop so that local people could access cheap, healthy food; organising clean-ups of the local green spaces (Hartcliffe Health and Environment Action Group); campaigning for better lighting, safer road crossings and other neighbourhood improvements (Riverside Youth and Community Centre, Snowhill); organising consultations on regeneration developments; campaigning for accessible public transport (e.g. South Gloucestershire Disability Equality Forum); developing and running a local community market so that people could access affordable food and goods locally; campaigning for a car share scheme – mostly only available in well-off areas (Lockleaze Neighbourhood Trust); and campaigning for tree planting and for safety measures in relation to the High Tension pylons that run through the estate (coinciding with high incidence of cancers and depression) (Lockleaze Environment Group). We were often blocked and constrained in these endeavours by those who made decisions on behalf of the community (councillors, local authority officers, development agencies) who seemed to lack the understanding or the will to engage in a participatory way (see Bell, 2008; Bell, 2020; Bell and Reed, 2021). Similarly, Harley and Scandrett (2019) note how, when community development in the UK initially began to engage with Local Agenda 21, which envisaged the involvement of civil society in transitioning to sustainability, insincere consultation processes served to hide the interests of the wealthy and powerful. Even so, working-class communities across the UK have continued to engage in environmental justice activities, within the constraints that are imposed on them.

Conclusion

'Environmentalism' has been framed in middle-class terms around organising in a specific way (illegal activism, green consumption), caring about a limited range of issues (climate change, biodiversity) and using messages which do not engage widely (consume less, look poor) (Bell, 2020). In the meantime, working-class people continue to care about and struggle for a healthy environment for all. Environmental and social justice would be better served by recognising and supporting these working-class struggles. XR and other mainstream environmental organisations have repeatedly apologised for their mistakes on inclusion (e.g. Extinction Rebellion, 2020) and conveyed their intention to do better. This is an important step, giving cause for optimism, though there is still a long road ahead. There is much more to know and learn about how to develop the necessary solidarities and I hope reading this chapter will lead to more exploration of the topic. Of course, the ideal scenario would be to end class-based inequality and the social divisions and justifications that go with it. Perhaps if we begin to mobilise across classes for a just transition to sustainability, we will gain enough empathy for each other that we will no longer be able to bear such a degenerate arrangement of society.

References

Abramson, P.R. and Inglehart, R. (1995) *Value Change in Global Perspective*. University of Michigan Press, Ann Arbor, MI.

Agyeman, J. (2001) 'Ethnic minorities in Britain: Short change, systematic indifference and sustainable development', *Journal of Environmental Policy and Planning*, vol 3, no 1, pp. 15–30.

Agyeman, J. (2002) 'Constructing environmental (in)justice: Transatlantic tales', *Environmental Politics*, vol 11, no 3, pp. 31–53.

Allen, K., Daro, V. and Holland, D.C. (2007) 'Becoming an environmental justice activist', in R. Sandler and P.C. Pezzullo (eds.), *Environmental Justice and Environmentalism: The Social Justice Challenge to the Environmental Movement*. Massachusetts Institute of Technology, Cambridge, MA.

Atkinson, W. (2015) *Class*. Polity, Cambridge.

Atkinson W. (2010) *Class, Individualization and Late Modernity: In Search of the Reflexive Worker,* Palgrave, Basingstoke.

Bell, K. (2008) 'Achieving environmental justice in the United Kingdom: A case study of Lockleaze', *Environmental Justice* vol 1, no 4, pp. 203–210.

Bell, K. (2014) *Achieving Environmental Justice*. Policy Press, Bristol.

Bell, K. (2020) *Working-Class Environmentalism: An Agenda for a Fair and Just Transition to Sustainability*. Palgrave, London.

Bell, K. and Bevan, G. (2021, forthcoming) Beyond Inclusion?: Perceptions of the Extent to Which Extinction Rebellion Speaks to, and for, Black, Asian and Minority Ethnic (BAME) and Working-Class Communities Rebellion in the United Kingdom'.

Bell, K. and Reed, M. (2021, forthcoming) 'The tree of participation: A new tool for engagers in participatory environmental decision-making', *Community Development Journal*.

Bennett, C. (2015) 'Green movement must escape its 'white, middle-class ghetto', says Friends of the Earth chief Craig Bennett', *The Independent*, 4 July

Bowden, S. and Tweedale, G. (2003) 'Mondays without dread: The trade union response to byssinosis in the Lancashire cotton industry in the twentieth century', *Social History of Medicine*, vol 16, no 1, pp. 79–95.

Bradley, H. (2014) 'Class descriptors or class relations? Thoughts towards a critique of Savage et al.', *Sociology*, vol 48, no 3, pp. 429–436.

Bufton, M.W. and Melling, J. (2005) '"A mere matter of rock": Organized labour, scientific evidence and British government schemes for compensation of silicosis and pneumoconiosis among coalminers, 1926–1940', *Medical History*, vol 49, no 2, pp. 155–178.

Bullard, R.D. (1983) 'Solid waste sites and the black Houston community', *Sociological Enquiry*, vol 53, no (2–3), pp. 273–288.

Bullard, R.D. (1990) *Dumping in Dixie: Race, Class, and Environmental Quality*. Westview Press, Boulder, CO.

Bullard, R.D. and Wright, B.H. (1992) 'The quest for environmental equity: Mobilizing the African-American community for social change', in R.E. Dunlap and A.G. Mertig (eds.), *American Environmentalism: The U.S. Environmental Movement, 1970–1990*. Taylor & Francis, New York.

Burningham, K. and Thrush, D. (2003) 'Experiencing environmental inequality: The everyday concerns of disadvantaged groups', *Housing Studies*, vol 18, no 4, pp517–536

Burningham, K. and Thrush, D. (2001) *Rainforests are a Long Way from Here: The Environmental Concerns of Disadvantaged Groups*. YPS for the Joseph Rowntree Foundation, York.

Burningham, K. and Thrush, D. (2003) 'Experiencing environmental inequality: The everyday concerns of disadvantaged groups', *Housing Studies*, vol 18, no 4, pp. 517–536.

Burt, J., Stewart, D., Preston, S. and Costley, T. (2013) 'Monitor of engagement with the natural environment survey (2009–2012): Difference in access to the natural environment between social groups within the adult English population', *Natural England Data Reports*, Number 003.

Capacity Global (2009) *Every Action Counts: The Diversity Report*. Capacity Global, London.

Clarke, L. and Sahin-Dikmen, M. (2018) 'Green transitions in the built environment in Europe. Workshop: What kind of green and just transition? University of Westminster 12 July.

CLASS (2017) 'CLASS on class'. Centre for Labour and Social Studies. https://player.fm/series/class-on-class

DEFRA (2011) *The Natural Choice: Securing the Value of Nature*. DEFRA, London.

Di Chiro, G. (1996) 'Nature as community: The convergence of environment and social justice', in W. Cronon (ed.), *Uncommon Ground: Rethinking the Human Place in Nature*, pp. 298–320. W.W. Norton, New York.

Dunlap R.E. and Mertig A.G. (1997) 'Global environmental concern: An anomaly for postmaterialism', *Social Science Quarterly*, vol 78, no 1, pp. 24–29.

Dunlap R.E. and York R. (2008) 'The globalization of environmental concern and the limits of the postmaterialist values explanation: Evidence from four multinational surveys', *Sociological Quarterly*, vol 49, no 3, pp 529–563.

Eisenberg-Guyot, J. and Prins, S.J. (2019) 'Relational social class, self-rated health, and mortality in the United States', *International Journal of Health Sciences*, vol 50, no 1, n.p.

Extinction Rebellion (2020) 'Statement on Extinction Rebellions relationship with the police', 1 July. https://extinctionrebellion.uk/2020/07/01/statement-on-extinction-rebellions-relationship-with-the-police/.

Fairburn, J., Butler, B. and Smith, G. (2009) 'Environmental justice in South Yorkshire: Locating social deprivation and poor environments using multiple indicators', *Local Environment*, vol 14, no 2, pp. 139–154.

First National People of Color (1991) 'Principles of environmental justice', *First Nation People of Color Environmental Leadership Summit*, Washington, DC.

Franzen, A. (2003) 'Environmental attitudes in international comparison: An analysis of the ISSP surveys 1993 and 2000', *Social Science Quarterly*, vol 84, no 2, pp. 297–308.

Franzen, A. and Meyer, R. (2010) 'Environmental attitudes in cross-national perspective: A multilevel analysis of the ISSP 1993 and 2000', *European Sociological Review*, vol 26, no 2, pp. 219–234.

Franzen, A. and Vogl, D. (2013) 'Two decades of measuring environmental attitudes: A comparative analysis of 33 countries', *Global Environmental Change*, vol 23, no 5, pp. 1001–1008.

Gelissen J. (2007) 'Explaining popular support for environmental protection: A multilevel analysis of 50 nations', *Environment and Behavior*, vol 39, no 3, pp. 392–415.

Goldthorpe, J.H. and Chan, T.W. (2007). 'Class and status: The conceptual distinction and its empirical relevance', *American Sociological Review*, vol 72, no 4, pp. 512–532.

Goodman, D. J. (2011) *Promoting Diversity and Social Justice: Educating People from Privileged Groups*, Sage, London.

Gottlieb, R. (1993) *Forcing the Spring. The Transformation of the American Environmental Movement*. Island Press, Washington.

Grossmann, M. and Creamer, E. (2017) 'Assessing diversity and inclusivity within the transition movement: An urban case study', *Environmental Politics*, vol 26, no 1, pp. 161–182.

Harley, A. and Scandrett, E. (eds.) (2019) *Environmental Justice, Popular Struggle and Community Development*. Policy Press, Bristol.

Harter, J.H. (2004) 'Environmental justice for whom? Class, new social movements and the environment: A case study of Greenpeace Canada, 1971–2000', *Labour/Le Travail*, vol 54, pp. 83–119.

Hinsliff, G. (2019) 'Extinction Rebellion has built up so much goodwill. It mustn't throw that away', *The Guardian*. 17 October.

Inglehart, R. (1977) *The Silent Revolution: Changing Values and Political Styles among Western Publics*. Princeton University Press, Princeton, NJ.

IPCC (2018) *Global Warming of 1.5 °C*. Intergovernmental Panel on Climate Change. www.ipcc.ch/report/sr15/.

IPPR (2018) *Beyond Eco – The Future of Fuel Poverty Support*. Institute for Public Policy Research, June.

Jackson, T. (2005) 'Motivating Sustainable Consumption a review of evidence on consumer behaviour and behavioural change a report to the Sustainable Development Research Network', Centre for Environmental Strategy, University of Surrey, January.

Josette, N. (2019) 'People of colour are the most impacted by climate change, yet Extinction Rebellion is erasing them from the conversation', *The Independent*, 21 April.

Knight, K.W. and Messer, B.L. (2012) 'Environmental concern in cross-national perspective: The effects of affluence, environmental degradation, and world society', *Social Science Quarterly*, vol 93, pp. 521–537.

Labor4sustainability (2017) 'Mission statement of the Labor Network for Sustainability'. www.labor4sustainability.org/wp-content/uploads/2017/03/LNS-Mission-and-Principles.pdf.

Laurison, D. and Friedman, S. (2016) 'The class pay gap in higher professional and managerial occupations', *American Sociological Review*, vol 81, pp. 668–695.

Lerner, S. (2012) *Sacrifice Zones: The Front Lines of Toxic Chemical Exposure in the United States*. The MIT Press, Cambridge, MA.

Loomis (2016) Towards a Working-Class Environmentalism, News Republic https://newrepublic.com/article/139132/towards-working-class-environmentalism

Lucas, K. (2012) 'Transport and social exclusion: Where are we now?', *Transport Policy*, vol 20, pp. 105–113.

MacPhee, K. (2014) 'Canadian working-class environmentalism 1965–1985', *Labour-Le Travail*, 74 (Fall 2014), pp. 123–149.

Malone, C. (2019) 'Extinction Rebellion is a "loopy middle-class doomsday cult"', *The Pledge*, 11th October. www.youtube.com/watch?v=aWXDo41mkGE.

Martinez-Alier, J. (2003) *Environmentalism of the Poor*. Edward Elgar, Basingstoke

Marquart-Pyatt, S. T. (2008) 'Are there similar sources of environmental concern? Comparing industrialized countries', *Social Science Quarterly*, vol 89, no 5, pp. 1312–1335.

Mastini, R., Kallis, G. and Hickel, J. (2021) 'A Green New Deal without growth?', *Ecological Economics,* vol 179, 106832.

Mcgarvey, D. (2017) *Poverty Safari*. Picador, London.

Mckenzie, L. (2015) *Getting by: Estates, Class and Culture in Austerity Britain*. Policy Press, Bristol.

Milojevic, A., Niedzwiedz, C.L., Pearce, J., Milner, J., MacKenzie, I.A., Doherty, R.M. and Wilkinson, P. (2017) 'Socioeconomic and urban-rural differentials in exposure to air pollution and mortality burden in England', *Environmental Health,* vol 16, no 104, pp. 1–10.

Mitchell, G. and Dorling, D. (2003) 'An environmental justice analysis of British air quality', *Environmental Planning A,* vol 35, pp. 909–929.

Morris, J. (2019) 'Extinction Rebellion activists occupy Smithfield Market and set up fruit and veg stalls', *The Standard,* 8 October.

Muntaner, C., Ng, E., Chung, H. and Prins, S.J. (2015) 'Two decades of Neo-Marxist class analysis and health inequalities: a critical reconstruction', *Social Theory and Health,* vol 13, no 3–4, pp. 267–287.

Narain, S. (2013) 'How to be or not to be year of environment', *Business Standard.* 13 January https://www.business-standard.com/article/opinion/sunita-narain-how-to-be-or-not-to-be-year-of-environment-111011000048_1.html

Nawrotzki, R.J. and Pampel, F.C. (2012) 'Cohort change and the diffusion of environmental concern: A cross-national analysis', *Population and Environment,* vol 35, no 1, pp. 1–25.

OECD (2018). *A Broken Social Elevator? How to Promote Social Mobility*, OECD Publishing, Paris. https://doi.org/10.1787/9789264301085-en

O'Neil, B. 020) Coal Miners vs Extinction Rebellion, Spiked 2nd March https://www.spiked-online.com/2020/03/02/coal-miners-vs-extinction-rebellion/

O'Neill, S. and Nicholson-Cole, S. (2009) '"Fear won't do it" Visual and iconic representations', *Science Communication,* vol 30, pp. 355–379.

Oxfam (2014) 'England's most deprived areas three times more likely to have been flooded than most well-off'. http://oxfamapps.org/media/8ncz5.

Pang, M., Meirelles, J., Moreau, V. and Binder, C. (2019) 'Urban carbon footprints: A consumption-based approach for Swiss households', *Environmental Research Communications,* vol 2, 011003, pp. 1–12.

Pellow, D.N. (2007) *Resisting Global Toxics: Transnational Movements for Environmental Justice.* The MIT Press, Cambridge, MA.

Pellow, D.N. (2018) *What Is Critical Environmental Justice?* Polity, Cambridge.

Power, A. and Elster, J. (2005) 'Environmental issues and human behaviour in low-income areas in the UK. CASEreports (31)', Centre for Analysis of Social Exclusion, London School of Economics and Political Science, London.

Pulido, L. (1998) 'Development of the 'People of Color' identity in the environmental justice movement of the Southwestern United States', *Socialist Review,* vol 26, no 3–4, pp. 145–180.

Pye, S.K., King, K. and Sturman, J. (2006) 'Air quality and social deprivation in the UK: An environmental inequalities analysis-final report to Defra', Contract rmp/2035.

Räthzel, N. and Uzzell, D. (2012) 'Mending the breach between labour and nature: Environmental engagements of trade unions and the North-South divide', *Interface: A Journal for and about Social Movements*, vol 4, no 2, pp 81–100.

Reay, D. (2017) *Miseducation: Inequality, Education, and the Working Classes.* Policy Press, Bristol.

Rector, J. (2014) 'Environmental justice at work: The UAW, the war on cancer, and the right to equal protection from toxic hazards in postwar America', *Journal of American History*, vol 101, no 2, pp. 480–502.

Richardson, E., Shortt, N. and Mitchell, R. (2010) 'The mechanism behind environmental inequality in Scotland: Which came first, the deprivation or the landfill?', *Environment and Planning A*, vol 42, pp. 223–240.

Rohrschneider, R. (1990) 'The roots of public-opinion toward new social-movements – An empirical-test of competing explanations', *American Journal of Political Science*, vol 34, no 1, pp. 1–30.

Satheesh, S. (2020) 'Moving beyond class: A critical review of labor-environmental conflicts from the Global South', *Sociology Compass* (in press).

Saunders, C., Doherty, B. and Hayes, G. (2020) 'A new climate movement? Extinction Rebellion's activists in profile', *CUSP Working Paper No 25*. Centre for the Understanding of Sustainable Prosperity, Guildford.

Scheidel, A., Del Bene, D., Liu, J., Navas, G., Mingorría, S., Demaria, F., Avila, S., Roy, B., Ertör, I., Temper, L., Martínez-Alier, J. (2020) 'Environmental conflicts and defenders: A global overview', *Global Environmental Change*, vol 63, 102104.

Shand-Baptiste, K. (2019) 'Opinion: Extinction Rebellion's treatment of class and race blocks its goal of climate justice', *The Independent*, 15 October.

Skeggs, B. (2004) *Class, Culture, Self.* Routledge, London.

Smith, D.H. (1997a) 'The international history of grassroots associations', *International Journal of Comparative Sociology*, vol 38, no 3–4, pp. 189–216.

Smith, D.H. (1997b) 'The rest of the nonprofit sector: Grassroots associations as the dark matter ignored in prevailing 'Flat-Earth' maps of the sector', *Nonprofit and Voluntary Sector Quarterly*, vol 26, no 2, pp. 114–131.

Social Mobility Commission (2017) *Social Mobility Policies between 1997 and 2017: Time for Change.* SMC, London.

Steffen et al. (2015). 'Planetary boundaries: Guiding human development on a changing planet', *Science*, vol. 347, no 6223.

Stevis, D., Kraus, D. and Morena, E. (2020) 'Introduction: The genealogy and contemporary politics of just transitions', in E. Morena, D. Krause and D. Stevis (eds.), *Just Transitions: Social Justice in the Shift Towards a Low-Carbon World.* Pluto Press, London.

Taylor, A. and Loopstra, R. (2016) *Too Poor to Eat: Food Insecurity in the UK.* Food Foundation, London.

Taylor, D.E. (1997) 'American environmentalism: The role of race, class and gender in shaping activism 1820–1995', *Race, Gender and Class*, vol 5, no 1, pp. 16–62.

Taylor, D.E. (2016) *The Rise of the American Conservation Movement: Power, Privilege, and Environmental Protection.* Duke University Press, Durham, NC.

Toft, M. and Friedman, S. (2020) 'Family wealth and the class ceiling: The propulsive power of the bank of mum and dad', *Sociology* (in press).

UCC CRJ (1987) 'Toxic wastes and race in the United States: A national report on the racial and socio-economic characteristics of communities with hazardous waste sites'. United Church of Christ Commission for Racial Justice, New York.

US GAO (1983) 'Siting of hazardous waste landfills and their correlation with racial and economic status of surrounding communities', General Accounting Office, GAO/RCED-83–168. Government Printing Office, Washington, DC.

Uyeki E.S. and Holland L.J. (2000) 'Diffusion of pro-environment attitudes?', *American Behavioral Scientist*, vol 43, no 4, pp. 646–662.

Uzzell, D., & Räthzel, N. (2009). Transforming environmental psychology, *Journal of Environmental Psychology*, 29, 340–350.

Veblen, T. (1994) [1899] *The Theory of the Leisure Class: An Economic Study in the Evolution of Institutions*. The Macmillan Company, New York.

Vosper, N. (2016) 'What makes me tired when organising with middle-class comrades', *The Guardian,* 8 June.

Walker, G. (2012) *Environmental Justice: Concepts, Evidence and Politics*. Routledge, London.

Walker, G., Burningham, K., Fielding, J., Smith, G., Thrush, D. and Fay, H. (2007) *Addressing Environmental Inequalities: Flood Risk*. Environment Agency, Bristol.

Walker, G. and Burningham, K. (2011) *Flood risk, vulnerability and environmental justice: Evidence and evaluation of inequality in a UK context*. https://core.ac.uk/download/pdf/9550048.pdf.

Walker, G., Mitchell, G., Fairburn, J. and Smith, G. (2003) 'Environmental quality and social deprivation. Phase II: National analysis of flood hazard, IPC industries and air quality', The Environment Agency, Bristol, R&D Project Record E2-067/1/PR1, p. 133.

Walkerdine, V. (2020) '"No-one listens to us": Post-truth, affect and Brexit', *Qualitative Research in Psychology*, vol 17, no 1, pp. 143–158.

Wheeler, B. (2004) 'Health-related environmental indices and environmental equity in England and Wales', *Environment and Planning A*, vol 36, no 5, pp. 803–822.

Wilkinson, R. and Pickett, K. (2010) *The Spirit Level: Why Greater Equality Makes Societies Stronger*. Bloomsbury Publishing, New York.

Wilkinson, R. and Pickett, K. (2019) *The Inner Level: How More Equal Societies Reduce Stress, Restore Sanity and Improve Everyone's Well-Being*. Penguin Press, London.

Wright, E.O. (2001) 'Foundations of class analysis: A Marxian perspective', in J. Baxter and M. Western (eds.), *Reconfigurations of Class and Gender*. Stanford University Press, Stanford, CA.

WTUL (1899) *Potteries Fund Register Lead poisoning in the potteries*, The Women's Trade Union League, TUC Library Collections, London Metropolitan University.

Wretched of the Earth (2019) 'An open letter to Extinction Rebellion', *Red Pepper,* 3 May.

6

ENVIRONMENTALISM AND LGBTQIA+ POLITICS AND ACTIVISM

Emma Foster

Introduction

The idea of the 'culture wars' in recent years has dominated research on right-wing populist politics in the United States and much of Europe. This research (e.g. see Inglehart and Norris, 2016) posits that the rise in right-wing populism is, in part, a backlash to the increasingly 'progressive' values that have become increasingly embedded in the West since the 1970s. Simply put, in the context of relative material security, areas where 'progressive' values have gained traction have experienced a cultural backlash from those (usually older, White men) who feel that their own values have been marginalised within their 'own countries'. The target of this cultural backlash are individuals, groups and sets of values deemed 'progressive' such as those identified with gender equality, racial equality, human rights, pro-migration, sexual rights and environmentalism. Arguably, this backlash, framed through the so-called culture wars is, at the time of writing, more acute than ever. Yet, queer ecofeminist Greta Gaard, writing in the mid-1990s, identified back then that the conservative right in the United States grouped these positions together as progressives, in the hope to see their 'collective annihilation' (1997, p. 114). In addition, Gaard notes that the irony is that these groups, so neatly pathologised as dangerous progressives by the right, are disunited and even antagonistic to one another. More than 20 years later, this lack of unity remains the case. With that in mind, this chapter begins from the premise that LGBTQIA+ and environmental activists, academics and proponents infrequently speak to one another, let alone merge projects or meaningfully collaborate.

This chapter focuses on these two 'progressive' groups, the LGBTQIA+ and environmental movements, who have rarely collaborated. The first section explores the barriers to an effective alliance between environmentalists and the LGBTQIA+ movement. Drawing from queer eco-critique, I highlight the ways in which, particularly mainstream, environmentalism is premised on heterosexual assumptions and perpetuates heteronormativity. The section that follows then explores the interesting and radical ways that these two movements have attempted to merge or forge a radical ecological politics. Here, the chapter focusses on the radical potential of a queer ecological politics and examines the academic literature in this field. I turn to look at how this queer sentiment overlaps with, and has manifested in, the environmental justice movement (EJM) and, more recently, among vegan activists seeking a comprehensive and decolonised ecological veganism.

However, before presenting the queer eco-critique of (mainstream) environmentalism, it is worth, at this juncture, to offer some definitional clarity about key groups and terms. While the LGBTQIA+ acronym is dynamic, evolving and contestable (Oakley, 2016), for the purpose of this chapter I am defining LGBTQIA+ as the overarching term describing marginalised sexualities and genders. This includes lesbian, gay, bisexual, transgender, transsexual, queer, questioning, intersex, a-sexual, a-gender and their allies (hence LGBTQIA+), as well as those who do not neatly correspond to the acronym, such as gender-queer, pansexual and polysexual. In other words, LGBTQIA+ encapsulates those sexual identities, relations and practices distinct from the heterosexual monogamous norm. The LGBTQIA+ movement, it follows, refers to the groups that fight for, advocate and promote diverse expressions of sexuality and gender, whether that is through the language of 'rights' (such as LGBTQIA+ rights) or more radical deconstructions of heteronormativity (where heteronormativity is the system that privileges and naturalises monogamous heterosexuality).

In the LGBTQIA+ acronym, Q is most often used to refer to queer and is a term some individuals identify with. This chapter uses the term 'queers', similar to the LGBTQIA+ acronym, to describe those individuals who reject gender and sexuality norms. However, it is important to note that the term is used within the acronym for those who do not identify as straight but equally find the terms 'gay', 'bisexual' or 'lesbian' delimiting. Those who identify as queer tend to recognise the categories of gay, bisexual and lesbian, as based on the hierarchy of sexualities upon which heteronormativity relies; being rooted in an arbitrary distinction between male and female; men and women. In other words, heterosexuality operates through binary gender distinction (male/female) and can only be dominant relative to those sexualities it is not – those sexualities that are considered deviant or abnormal. Furthermore, if doubt is cast over the categories 'men/male' and 'women/female', the gender of the desiring and desired can no longer be used as the factor determining sexual identity. Indeed, while Q is a rather complicated identifying letter under the LGBTQIA+ umbrella term, the other reason I draw particular attention to it here is because 'queer' also refers to an academic theory, as in *queer theory*. Queer theory is an academic theory that attempts to locate, denaturalise and generally critique the systems of heteronormativity that underpin politics, economics and society (Sedgewick, 1993). As such, queer theory has been at the centre of the relatively sparse approaches that seek to bridge sexuality and ecology. These approaches are often subsumed under the umbrella term of queer ecology (e.g. see Gaard, 1997; Sandilands, 2002; Mortimer-Sandilands, 2005, Mortimer-Sandilands and Erickson, 2010; Morton, 2007).

LGBTQIA+ against nature? The disconnect between mainstream environmentalism and the LGBTQIA+ movement

While environmentalism has many different strands, in general, nature is the object of environmentalism. What I mean by this is that nature is often the thing that environmentalists seek to save, conserve, preserve, protect and/or liberate. It is the object of their activism in a similar way as women are the focus of feminism and the working classes are the focus of socialism. As such, a coherent understanding of nature is deemed crucial for any environmental project. This means that the environmental movement, perhaps more explicitly than any other movement, is fundamentally informed by how it defines nature, and subsequently what is natural or unnatural. It is this compulsion to define nature/natural/unnatural that arguably inhibits a meaningful dialogue between environmentalists and the LGBTQIA+ movement.

Queer ecologist (and ecofeminist) Greta Gaard, who I introduced in the opening paragraph of this chapter, highlights in her seminal article 'Toward a Queer Ecofeminism' (1997) that the dichotomy of natural and unnatural works to reinforce heterosexual privilege and subordinate non-normative sexualities. This is because heterosexuality is commonly understood to be the only natural form of sexuality and is therefore the only form that is genuinely permissible. The reason heterosexuality is considered natural is because of the role it plays in procreation which is seen as commensurate with the supposedly natural evolutionary drive to bequeath genes through the continuation of the species. Non-normative sexual practices, it follows, seen as non-procreative were/are considered deviant as they disrupt these 'natural' drives and impulses and the natural/evolutionary order (Mortimer-Sandilands, 2005). This natural/unnatural logic works to devalue and subordinate non-normative sexual identities and practices, and has historically legitimised criminalisation, medicalisation, exclusion and marginalisation of those who do not subscribe to (procreative) heterosexuality (Bell, 2010; Gosine, 2010; Hogan, 2010).

Attaching naturalness to heterosexuality and unnaturalness to sexualities that depart from heterosexuality has worked to devalue and subordinate the latter, legitimising legal and medical scrutiny and the policing of those who would in contemporary parlance identify as LGBTQIA+. That said, it is important to note that there is a fundamental contradiction in this thinking. Given that the natural/unnatural logic has been operationalised to subordinate queers of all kinds, one would assume that what is natural, and therefore nature, would be afforded considerable value in dominant culture. However, as environmental degradation testifies, it isn't. As Gaard so astutely notes:

> On the one hand, from a queer perspective, we learn that the dominant culture charges queers with transgressing the natural order, which in turn

implies that nature is valued and must be obeyed. On the other hand, from an ecofeminist perspective, we learn that Western culture has constructed nature as a force that must be dominated if culture is to prevail.

(Gaard 1997, p. 143)

She goes on to argue that this contradiction exposes that the nature that queers are being encouraged to sign up to is reduced to heterosexuality and not that which is considered the natural world, in all of its rich diversity. Nonetheless, despite this paradox, which I shall return to in the following section, it remains that nature (or constructions of naturalness) has been used as a tool to maintain the marginalisation and subordination of non-normative sexualities. Often premised in religious discourses of same-sex sexualities amounting to a 'crime against nature', this sentiment continues to hold much weight. One good example, where the relationship between non-normative sexualities and the destruction of nature was constructed as equivalent, came in Pope Benedict XVI's 2008 Christmas greeting to the Vatican staff. Here he took the opportunity to talk about human ecology as that of the nature of man and woman. He continued to explain how ignoring that ecology was tantamount to the destruction of the rainforests (Hogan, 2010, p. 4966). In this context, it is unsurprising that many LGBTQIA+ people have been reluctant to subscribe to those elements of an environmental movement that hold a firm conceptualisation of nature as heterosexuality and, subsequently, valorise it. So much so that queer politics and queer theory, rather than finding affinity with ecologism, has actually demonstrated a tendency towards bio-phobia (Garrard, 2010).

While understandably suspicious of 'nature' and how it is wielded at the expense of those categorised as queer, numerous examples of LGBTQIA+ people being constructed in environmentally negative ways, most notably as akin to pollution, are common in scholarly queer eco-critique. For example, Katie Hogan (2010) briefly reviews the novel *Eight Bullets: One Woman's Story of Surviving Anti-Gay Violence* by Claudia Brenner and Hannah Ashley (1995), highlighting how non-conforming gender and non-normative sexual identities are excluded, sometimes forcefully, from rural spaces. This echoes the anthropological/sociological work done by Valentine (2002) and Bell and Valentine (1995) exploring the self or forced exclusion of LGBTQIA+ people from rural communities in favour of urban spaces. This displacement of LGBTQIA people from rural to urban spaces is because the city is seen to be more tolerant of diverse expressions of gender and sexuality. Indeed, the attempted clearance of non-normative sexual identities from 'natural' spaces is well documented in the eco-critical literature. Gosine (2010), for example, draws parallels between the exclusion of non-White indigenous communities from national parks and gay (men) from park spaces used for park sex. Here both non-White indigenous communities and gay men are seen as polluting, contaminating and damaging natural spaces. Regarding indigenous communities, indigenous peoples have been constructed as undermining the 'purity' of the ostensibly 'pristine nature' of the wilderness (Stanley, 2020, pp. 244–245; Hay, 2019) and displaced from these areas as a consequence (Bray and Velazquez, 2009). Gay men (meeting for sex) in urban parkland have, similarly, been seen to undermine the idea of 'family friendly' heterosexual park space purposed as respite from the degeneracy of the city (Mortimer-Sandilands, 2005).

Indeed, many environmental movements are premised on racialised, gendered and heterosexualised norms. For example, according to Mortimer-Sandilands (2005), the contemporary US environmental organisation, the Sierra Club, set up by John Muir in the late 19th century, was a key supporter of the national park movement, which sought to 'preserve' and cultivate areas of wilderness for human recreation. The justification to set up national parks was in itself racist (and heterosexist) as the advocates thought these spaces would offer 'clean spaces' for White people as refuge from industrialised areas that were increasingly inhabited by non-European immigrants (Mortimer-Sandilands, 2005, p. 5). Furthermore, through this initiative, many indigenous people were displaced to produce these 'clean' recreational spaces. Also embedded in the justification to set up national parks was the growing anxiety, at the time, of women gaining more political and economic power. In an attempt to assert masculinity in the context of these shifting gender dynamics, the parks were designed for men to engage in 'masculine' recreation like hunting and fishing, thereby superimposing these natural spaces with a very masculinised, 'tame and conquer', approach to the environment (Mortimer-Sandilands, 2005, p. 6; Stanley, 2020, p. 244). According to Stanley (2020), in reference to national parks, the idea of the outdoors is rooted in gender normativities that work to determine and regulate who can legitimately enjoy those spaces. Focusing on women hikers, Stanley notes that even when women are represented participating in outdoor activities, they tend to meet certain heterosexualised expectations, for example, by being 'accompanied by straight male partners and/or children' (Stanley, 2020, p. 249). Indeed, the reproduction of naturalised gender norms in and of itself is part and parcel of normative heterosexuality, with the idea of active men and passive women fulfilling their apparently naturally ordained reproductive imperative. In all, as Mortimer-Sandilands astutely concludes, national parks, as an expression of an early environmental initiative, were:

> [B]orn from a gendered and racialized view of nature, and were also used to impose gendered and racialized relations on nature. In turn, parks supported and extended racialized and class ideals of masculinity, and literally erased aboriginal peoples from the landscape, with fairly disastrous results for all concerned, including nature.
>
> *(Mortimer-Sandilands, 2005, p. 6)*

Arguably, then, expunging those people considered as contaminants, including LGBTQIA+ identifying individuals, from the so-called natural or wilderness spaces, is not only unjust but also has negative consequences for nature. With that in mind, the following section explores how a radical and just environmental politics can emerge from the eco-critique outlined above.

Queer eco-critique and radical ecology

The paradox identified earlier in this chapter, where nature is conflated with heterosexuality and presented as a benchmark of propriety while simultaneously being devalued, is one that is of interest to (eco)feminists and queer scholars alike (Gaard, 1997; Plumwood, 2002; Alaimo, 2008). However, it is through acknowledging this devaluation of nature that the interconnected and multiple ways that subordination, oppression and exploitation work become apparent. Through nature, sexuality, race and gender are made explicit and, as a result, solidarities can arise. Ecological feminists such as Karen Warren (1997) and Val Plumwood (2002) have gone to great lengths to demonstrate the pernicious inequalities that follow from Cartesian logics. They argue that contemporary societies continue to be based on the schema set out by the 17th-century philosopher Rene Descartes. Descartes argued that the mind and body were paired (as a dualism) but in opposition (distinct from one another). In the formula, the mind is related to consciousness and thinking, while the body is merely matter. As such, the mind when constructed as distinct from the body is simultaneously constructed as superior to the body. However, the oppositional dualisms do not end there as this mind/body split is interlinked and generative of multiple other oppositional, hierarchical dualisms. These include culture/nature, masculine/feminine, White/non-White, human/animal, reason/emotion and, as Gaard (1997) suggests, could also be expanded to include normative/queer. In this series of logics, the first term in the binary is privileged over the second and all the first terms (mind, culture, masculine, White, human, reason and normative) are perceived as related and mutually reinforcing. Similarly all the second terms (body, nature, feminine, non-White, animal, emotion and queer) are also perceived as interlinked and mutually reinforcing. For example, this is well illustrated when we think of the cultural currency afforded to feminised symbolic representations of nature, like mother earth (linking women and nature) or the constructions of indigenous people, people of colour, women and queers as closer to nature and at the mercy of their instincts, drives and emotions.

Indeed, the grouping together of indigenous people, people of colour, women and queers as 'closer to nature' has been used to justify their subordination and their exclusion from rights and (full) citizenship. At the same time, however, as an act of resistance these nexuses have also been repurposed by some within the environmental movement to demonstrate an affinity that leads to those individuals and groups (indigenous people, women, etc.) as having a superior understanding of the natural world and how to protect, heal or liberate it. The idea that nature, women, people of colour, indigenous people and queers share an affinity is a potential springboard for coalition. However, if these links are naturalised, even in the affirmative, this alliance runs the risk of reinforcing the very logics that have legitimised oppression. This is because, in reinforcing this ostensible closeness to nature, no attempt is made to challenge the binary thinking, the Cartesian dualism itself, which underpins the stratification of each term (Plumwood, 2002). As such, queer ecologists in particular recognise the importance of acknowledging that, rather than there being a natural or biological disposition that links certain individuals with nature, it is the discursive construction of this series of nexuses that draws nature, women, indigenous people, people of colour and queers together and entwines their oppressions. It is the complex inter-relationship of multiple oppressions (sexism, racism, heterosexism and speciesism), which so often operationalises the idea of nature, rather than any essential connection, that offers the potential to coordinate a comprehensive anti-oppression politics. This anti-oppression politics would rely on the drawing together of environmentalism with LGBTQIA+ movements, but simultaneously necessitates the inclusion of all those also fighting against racism, sexism and speciesism.

For queer ecologists, then, centralising a critique of nature presents an 'opportunity to build theoretical overlaps and an opportunity for activist coalitions among seemingly disparate groups and communities' (Hogan, 2010, p. 4989). Furthermore, in acknowledging the ideological work done via dominant conceptualisations of nature, eco-critique can also be used as a platform to challenge dominant conceptualisations and reconstruct more inclusive and just environmentalisms. Again, as Hogan insightfully notes:

> For queers, as for many racial, ethnic, gender, and religious minorities, nature is not a hiding place from ideology but often its location, and using nature as a resource or tool for ecocritique is a way to broaden what 'counts' as environmentalism.
>
> *(Hogan, 2010, p. 5087)*

Yet while queer ecologists recognise the affiliation between oppressions, and attempt to draw sexual, racial, ethnic, gender and religious minorities together in a politics of anti-oppression that also seeks to emancipate 'nature', their work tends to foreground and visibilise the relationship between sex, sexuality and nature. Furthermore, analyses have extended beyond exploring how queers (along with women and people of colour), supposedly unable to manage their emotions and urges, are constructed as closer to nature. Analyses now examine how nature has been sexualised and how the process of that eroticisation has justified the exploitation of the natural world. In other words, analyses do not just look at how people are oppressed through their naturalisation but how nature is rendered inferior or suspect through its eroticisation (see Gaard, 1997; Sandilands, 2002).

Returning to the Cartesian dualism discussed above, body, nature, women, people of colour, queers and animals are all constructed within a matrix of meaning that links them with each other. Indeed, many ecofeminists (see, e.g., Merchant, 1980; Collard and Contrucci, 1989; Seagar, 1993) have long argued that the feminisation of nature has legitimised its subordination and exploitation. As ecofeminist Carolyn Merchant argued so eloquently in *The Death of Nature* (1980), 'nature cast in the female gender, when stripped of activity and rendered passive, could be dominated by science, technology, and capitalist production' (Merchant, 2006, p. 514). In a similar vein, many queer ecologists have sought to demonstrate how the *eroticisation* of nature, which is so intimately linked with its feminisation, has also led to legitimising attempts to control and conquer what is perceived to be the natural world. Gaard demonstrates this well, highlighting how the feminisation and eroticisation of nature, simultaneous to the masculinisation of culture, works to present nature and culture as inter-related through compulsory heterosexuality (Gaard, 1993). She goes on to argue, in 'Towards a Queer Ecofeminism', that this eroticisation was crucial to the Christian missionary and colonial projects, noting that:

> Colonization becomes an act of the nationalist self asserting identity and definition over and against the other – culture over and against nature, masculine over and against feminine, reason over and against the erotic. The metaphoric 'thrust' of colonialism has been described as the rape of indigenous people and of nature because there is a structural – not experiential – similarity between the two operations, though colonization regularly includes rape.
>
> *(Gaard, 1997, p. 131)*

Gaard (1997) goes on to explain that in order to rationalise the seizing of land and people, colonial missions needed to demonstrate that their wars were just; that they were doing God's work rather than merely increasing the colonial powers' resources and control. This legitimisation process was very much filtered through the lens of sexuality where sexual and non-normative gender behaviour in indigenous populations was constructed as 'ungodly' and evidence that a 'civilising' Christian intervention was necessary. As such, same-sex desire, as well as transgenderism and sex acts between opposite sex individuals that departed from normative gender scripts, was used as a justification for colonisation. Subsequently, these diverse ways of being were pathologised, marginalised and replaced by Christian moral codes based in a fear of the erotic, based on erotophobia. Moreover, non-normative sexual behaviours practised by indigenous people were presented as animalistic, demonstrative of their supposed closeness to nature and distinction or separation from God. This demonstrates the way in which the erotic is placed within the realm of nature and how nature, here through animalism, is eroticised. As a result, erotophobia, which persists in Western liberal traditions, is operational in justifying the exploitation and subordination of nature – as a realm that needs taming and civilising to ensure its adherence to the normative heterosexual imperative. Unsurprisingly, this is of interest to queer ecologists who see erotophobia as a tool that has legitimised the violent subordination of queers, alongside people of colour, women, animals and, indeed, nature (see also Gaard, 2015; Gosine, 2010; Hatfield and Dionne, 2014). Acknowledging this as a simultaneous and mutually reinforcing process is crucial to attempts to synthesis LGBTQIA+ and environmental projects and movements, as explored in the next section.

LGBTQIA+ and environmental action/activism

For over 30 years many countries have seen the emergence, and even establishment, of green political parties in parliamentary politics. More recently, typified by Greta Thunberg's highly publicised climate strikes and the often controversial activism of international environmental groups like Extinction Rebellion (XR), we have witnessed an increasingly mobilised and active environmental movement, politicising climate change and environmental degradation. However, these green parties and environmental organisations and campaigns, while explicitly or implicitly supportive of the LGBTQIA+ movement and LGBTQIA+ equality and rights, do not express any synthesis between each project. For example, while the *Green Party of England and Wales* has arguably embedded commitments to LGBTQIA+ (or LGBTIQ as they would have it) equality (The Green Party, 2015), their position is not premised on the inter-relationship between the exploitation and degradation of nature and the subordination and oppression experienced by the LGBTQIA+ community. In other words, they present LGBTQIA+ equality as a policy commitment but it is not plugged into their wider environmental philosophy. Similarly, with groups like XR, there is a clear and direct emphasis on inclusivity, especially in their 6th principle: '[w]e welcome everyone and every part of everyone' (Extinction Rebellion, 2011). Again, this is not premised on an explicit commitment to LGBTQIA+ politics, but rather a commitment to inclusivity, which presumably embraces those who identify as, or affiliate with, LGBTQIA+. As such, when thinking about well-known green organisations and the LGBTQIA+ movement, is seems that both projects are seen as distinct, even if they are not positioned in opposition to one another. Well-known environmental organisations make no attempt to amalgamate green and sexual minority politics as the premise for their environmental ethic and, therefore, remain far removed from the queer ecologist perspective outlined above.

That said, it is important to note that while environmental and LGBTQIA+ projects are not synthesised in any meaningful way, there is certainly crossover of activists between groups. Also, and relatedly, there seems to be a willingness to corroborate in terms of sharing activist tactics and experiences. For example, *Greenpeace* released an article in February 2020, as part of LGBTQ+ History Month, that reviewed 'three incredible queer rights movements and their key figures, to reflect on what they have taught today's environmental activists' (Greenpeace, 2020, n.p.). Similarly, but from a slightly different perspective, in a recent blog entitled 'Is the Extinction Rebellion Movement Queer Inclusive?', Ash Kotac (2020), an XR and queer activist, argues in support of XR's inclusivity and, referencing what is termed rainbow racism, argues that it is actually queer groups that have become less progressive in recent years. He subsequently concludes by compelling queer activists to learn from movements like XR. Indeed, this sentiment is actually enshrined in XR's values, declaring that '[w]e value reflecting and learning: Following a cycle of action, learning and planning for more action. Learning from other movements and contexts as well as our own experiences' (XR, 2011, np). Overall then, there seems to be an appetite for cross-collaboration between the LGBTQIA+ and environmental movements that, although scarcely amounting to a greening of queer politics or queering of green politics, potentially reveals avenues towards a more coherent consolidation of anti-oppression activism.

While efforts towards an authentically queer ecology remain unrealised in the most recognised environmental organisations and parties, there is emerging evidence that queers have found traction for coalition and activism through the language of environmental justice. Environmental justice scholarship (e.g. see Bell, 2014) has tended to explore the ways that environmental degradation is experienced differently depending on class, race, ethnicity and gender. This scholarship highlights how poor people, people of colour, immigrants and women are more likely to experience the negative impacts of environmental degradation and less likely to experience environmental self-determination. Queer ecologists, inspired by the EJM and related scholarship, extend this approach to consider sexual minorities as well.

Katie Hogan, for example, in her chapter 'Undoing Nature: Coalition building as queer environmentalism', argues that queer eco-critique is 'a powerful contribution to theory and activism' (Hogan, 2010, p. 4989). This is because queer eco-critique interrogates the very concept that environmentalism is built upon, namely, nature, exposing the ideologies that operate through its construction. The implications for the EJM are profound and multiple. First, a queer environmental justice highlights the inter-relationship between ideologies of nature and the environmental injustices experienced by queer people, poor people, people of colour, women and so on. For example, as noted above, Gosine (2010) and Mortimer-Sandilands (2005) have argued how current conceptualisations of nature operate through racist and heterosexist logics to construct people of colour and sexual minorities as against nature. In so doing, they are regarded as polluting natural space. As a result, these apparently contaminating individuals are discouraged from inhabiting or even visiting rural areas. This exclusion is reinforced through a myriad of practices including subtle forms of marginalisation and micro-aggressions, exclusion based on property prices and even overt violence. This means people of colour and queers are more likely to live, and be visible, in urban spaces (Valentine, 2002) where environmental degradation has a greater impact on public health. Therefore, marginalised people, including queers, tend to be subjected to environmental injustice. This exclusion of people of colour and queers from rural areas on the grounds that they are 'against nature' and subsequently 'don't belong' not only resonates with the EJM's own critique, but also works as a catalyst for a broader coalition of those who bear the brunt of ideologies of nature. As Hogan notes, 'nature is an opportunity to build theoretical overlaps and an opportunity for activist coalitions among seemingly disparate groups and communities' (Hogan, 2010, p. 4989). I would also add that exposing the political work done through existing conceptualisations of nature presents an opportunity to reimagine natures in line with wider social justice objectives.

Overlapping with queer environmental justice work, which tends to focus on the links between heterosexism and racism in access to environmental goods and self-determination, there is an increasingly animated politics, albeit mostly online and in the form of zines and pamphlets, linking multiple intersectional experiences of oppression with veganism and environmentalism/ecologism. Here the role of speciesism is foregrounded as inextricably linked to racism (Feliz Brueck, 2019; Feliz Brueck and McNeill, 2020; Ko and Ko, 2017) and heterosexism or erotophobia

(Feliz Brueck and McNeill, 2020; Hamilton, 2019; Simonsen, 2012). This emerging literature tends to primarily focus on the relationship between humans and non-human animals and how this inter-relates with a myriad of other oppressions mapped onto existing social cleavages. In particular, equivalence is drawn between the justifications for the mistreatment of other-than-human animals, people of colour and queers. For example, in Feliz Brueck and McNeill's (2020) edited collection of short essays from queer vegan activists, the parallels drawn between environmental degradation, the exploitation of other-than-human animals and the subordination of people of colour, queers and women are rendered stark, when one contributor notes:

> Climate catastrophe is, in reality, an economic, political, neo-colonial, social justice, human rights life-and-death issue, and in conjunction with the exploitation of non-human animals, it's one more of the many ways white supremacy harms and kills marginalised people.
>
> *(Chris H. 2020, p. 571)*

This collection of insights from queer vegan activists in Feliz Brueck and McNeill's *Queer and Trans Voices: Achieving Liberation through Consistent Anti-Oppression*, problematises the whiteness, straightness and cis-ness of mainstream veganism and instead seeks to expose the mutually reinforcing relationship between a variety of *isms*. In Feliz Brueck's own contribution to the collection, she argues that:

> [M]ainstream veganism prioritizes the violence perpetrated against non-human animals above all other marginalised peoples without recognizing how our speciesist society also harms the environment, slaughterhouse workers (many of whom are Black, Brown, Indigenous People of Color and/or immigrants in the US), low income communities who live near factory farms and suffer from health issues connected to them, etc., and how the most privileged have historically harmed now vulnerable communities, including LGBTQIA+ people, while proclaiming these same groups are now responsible ethically to undo the harm against [nonhuman animals as] this one chosen group.
>
> *(Brueck, 2020, p. 331)*

The general thrust of this collection of papers, premised on problematising White veganism, seeks to present more radical forms of veganism informed by shared experiences of mutually enforcing exploitations. The authors go into detail about the relationship between animalisation and oppression arguing that 'white people, are taught to "dehumanize" and "animalize" every living being who isn't admitted into the tiny category of white people' (Chris H. 2020, p. 560). They see this animalisation as indicative of the disrespect 'white people' have shown towards nature and as operational in legitimisation of numerous inter-related oppressions (of the environment, other-than-human animals, people of colour, queers and women – in other words, everything that does not fit in with the 'tiny category of white people', to borrow Chris H.'s words).

Furthermore, Feliz Brueck and McNeill's collection, aligning with previous work that looks at the queer potential of a vegan diet such as Rasmus Simonsen's (2012) now famous 'A Queer Vegan Manifesto' and C. Lou Hamilton's (2019) more recent 'Veganism, Sex and Politics', demonstrate the challenge veganism poses to normative relations. Following Carol Adam's assertion that food is gendered and sexualised through various myths, not least the conceptualisation of the hunter gatherer, Simonsen, Hamilton and many of the authors featured in Brueck and McNeill identify the way that veganism challenges gender norms as well as the related norms that work to devalue nature. For instance, in Feliz Brueck and McNeill, Constantine, discussing the resistance felt by queers and animal liberation activists, notes:

> Like queer people, animal liberation activists face great resistance from society because their assertion of our kinship with non-humans challenges the carefully constructed dichotomies that justify society's manner of existence.
>
> *(Constantine, 2020, p. 722)*

As such, veganism can be read as a queer act in so far as it disrupts the normative behaviours and assumptions linked to gender, sexual and species relations. Similar to Greta Gaard's work on queer ecofeminism, these activists and activist-academics working towards a queer vegan ethic recognise the mistreatment of animals to be another oppression inextricably linked to the oppression faced by people of colour, queers, women, immigrants, other-than-human animals, the environment and so on. At the same time, they recognise that you cannot fight any of these issues in isolation as they are co-dependent. As a result, they call for a coherent anti-oppression politics that combines and arguably strengthens the challenge to existing power relations, and consequently works towards a more harmonious ecology.

Contribution

As noted above, dominant ideologies of nature, which underpin mainstream environmentalism, do political work and have legitimised the mistreatment of some human-animals, including those who would identify as, or affiliate with, LGBTQIA+. For queer ecologists, recognising nature as ideological and interrogating the ways in which this ideology operates in marginalising sexual minorities (as well as all 'others' who are constructed as 'closer to nature' and, sometimes paradoxically as 'unnatural' in the same breath) is the basis for a new radical ecological movement. In other words, eco-critique and interrogating these dominant understandings of nature works as a springboard for a new environmental politics (Hogan, 2010).

For example, for me (Foster, 2016) as well as others (e.g. see Mortimer-Sandilands, 2005), the promise of a queer ecology, in problematising the reproductive imperative crucial to mainstream environmental discourses of future generations, yields much potential for a radical environmental politics. Mainstream environmental governance and activity continues to culminate around ideas of sustainable development, where sustainable development evokes an image of future generations, asking individuals to protect the environment for their children and their children's children. This call is problematic as, by implication, it is limited to only those who have had, or who plan to have, children, thereby invisibilising care for the environment that is not rooted in having a genetic stake in the future (Ensor, 2012). Second, and relatedly, it is based on a very instrumental view of the environment, where individuals are encouraged to protect only that which their biological human offspring depend. This presents a very precarious environmental ethic where only the aspects of nature perceived to be useful to sustain the life of human offspring generally, and one's own genetic futurity specifically, are considered worthy of conservation or protection. A queer ecology perspective, on the other hand, disrupts the heteronormative premise of future generations and blurs boundaries between other-than-human-animals, human-animals, nature, culture and so on. In blurring these boundaries a more radical and expansive environmental politics can emerge that does not reduce the impetus for protecting the planet only as far as those aspects of nature are considered crucial to the futurity of one's offspring. Instead, the impetus may be found in acknowledging that multiple oppressions, marked by sexuality, gender, race, class, species and so on, are mutually reinforcing and co-dependent and, as a result, need to be tackled synchronously. Acknowledging this affinity may well result in an ethic of care that goes beyond our biological kin, contemporary and in the future, towards an inclusive ethic of care that transcends societal and species categories.

Concluding thoughts and recommendations: politics and policy

Following the logic of the above, in relation to practical politics, one should be heartened by the commitment to LGBTQIA+ rights within established green parties in the UK (and elsewhere) as well as the 'cross-fertilisation' occurring between queer and environmental movements, such as XR. However, as noted above, this does not go far enough as these movements fail to recognise the mutually reinforcing character of the oppressions they are facing and, subsequently, the imperative to sincerely combine efforts into a broader anti-oppression movement. Simply, it is a compelling argument that queer and environmental campaigns would benefit from acknowledging that violence towards queers is inextricably linked to the violence against nature, people of colour, women, immigrants, indigenous people and other-than-human animals.

More promising still are the recent efforts (Simonsen, 2012; Ko and Ko, 2017; Hamilton, 2019; Feliz Brueck, 2019; Feliz Brueck and McNeill, 2020) to demonstrate through blogs, books and online activism, this inter-relationship between multiple and intersectional oppressions. This work has emerged through debates around animal rights/liberation, racial politics and trans and queer politics and seeks to promote a wider politics of anti-oppression. Indeed, this emerging group of (often academic) activists believe that a movement that tackles oppression avoids the divisions between groups that have been constructed through mainstream discourses. Those divisions, which are reproduced through the dominant narratives, undermine solidarities and compromise efforts as different groups, marked by race, gender and sexuality, among other things, compete for acceptance and rights within the existing system. A better and arguably more effective approach would be to acknowledge the inter-dependence of oppressions and to tackle oppression itself, as one thing.

In addition, in contrast to erotophobia as it relates to the fear of and degradation of nature (Gaard, 1997, 2015; Hatfield and Dionne, 2014), some queer ecologists explicitly promote encouraging sex-positive and intimate relations with nature, seeing this as another site for a less precarious and more inspiring ecological politics (Sandilands, 2002; Alaimo, 2019; Whitworth, 2019). Indeed, Whitworth (2019) in her article 'Goodbye Gauley Mountain, hello eco-camp', exploring ecosexuality and the sex ecology movement, highlights how a more sensual, intimate or sexual engagement with that which is culturally intelligible as nature could pave the way for a better environmental ethic. Whitworth argues that ecosexuality and sex ecology, while not completely blurring the boundaries of culture and nature that Plumwood (2002) so eloquently criticised, represent a challenge to human exceptionalism and may 'provide a mode of thinking and being that demonstrates a certain openness to the world' (Whitworth, 2019, p. 88). Similarly, Alaimo (2019, p. 409) argues that 'intimate relationality with other species, whether they be immediate or mediated, literal or speculative, practical or aesthetic' may promote a firmer commitment to sustain species threatened by extinction. As such, queer ecological approaches open up new ways of thinking about 'nature' that have long been prohibited through logics of erotophobia and that may well have the potential to encourage a radical ecological politics.

While there is a case for LGBTQIA+ and environmental movements and organisations to merge their efforts alongside other anti-oppression groups, and while queer ways of thinking about nature may inspire creative sex-positive environmental movements, it is more difficult to say how policy itself should be shaped. However, one in-road might be for established environmental organisations to acknowledge the ways in which mainstream environmental policy is premised on heteronormativity. Policy initiatives might then be more directed towards promoting attitudinal changes where 'nature' is worth emancipating/protecting in its own right and not just a source of environmental goods and services for the survival of contemporary human-animals and their biological offspring (Foster, 2014, 2016). In addition, taking cues from the EJM, on a practical and immediate basis, decision makers should base policy on the assumption that natural degradation disproportionately impacts already marginalised groups, including queers, and that everyone, including queers, has the right to environmental (as well as political, economic and cultural) self-determination.

References

Alaimo, S. (2008) 'Trans-corporeal feminisms and the ethical space of nature', *Material Feminisms*, vol 25, no 2, pp. 237–264.

Alaimo, S. (2019) 'Wanting all the species to be: Extinction, environmental visions, and intimate aesthetics', *Australian Feminist Studies*, vol, 34, no 102, pp. 398–412.

Bell, D. and Valentine G. (1995) 'Queer country: Rural lesbian and gay lives', *Journal of Rural Studies*, vol 11, no 2, pp. 113–122.

Bell, D. (2010) 'Queernaturecultures', in C. Mortimer-Sandilands and B. Erickson (eds.), *Queer Ecologies: Sex, Nature, Politics, Desire*. Kindle, Indiana University Press, Indianapolis, IN.

Bell, K. (2014) *Achieving Environmental Justice: A Cross-National Analysis*. Policy Press, Bristol, UK.

Bray, D.B. and Velazquez, A. (2009) 'From displacement-based conservation to place-based conservation', *Conservation and Society*, vol 7, no 1, pp. 11–14.

Brenner, C. and Ashley, H. (1995) *Eight Bullets: One Woman's Story of Surviving Anti-Gay Violence*. Firebrand Books, Ithaca, NY.

Collard, A. and Contrucci, J. (1989) *Rape of the Wild: Man's Violence against Animals and the Earth*. Indiana University Press, Indianapolis, IN.

Constantine, M. (2020) 'Oppressive dichotomies: Fighting for animals with queer liberation', in J. Feliz Brueck and Z. McNeill (eds.), *Queer and Trans Voices: Achieving Liberation through Consistent Anti-Oppression*. Sanctuary Publishers, Kindle.

Ensor, S. (2012) 'Spinster ecology: Rachel Carson, Sarah Orne Jewett, and Nonreproductive Futurity', *American Literature*, vol 84, no 2, pp. 409–435.

Extinction Rebellion (XR) (2011) 'About us'. https://rebellion.global/about-us/ (accessed 02 August 2020).

Feliz Brueck, J. (2019) *Veganism of Color: Decentring whiteness in human and non-human liberation*. Sanctuary Publishers, Kindle.

Feliz Brueck, J. and McNeill, Z. (2020) 'Introduction', in J. Feliz Brueck and Z. McNeill (eds.), *Queer and Trans Voices: Achieving Liberation through Consistent Anti-Oppression*. Sanctuary Publishers, Kindle.

Foster, E.A. (2014) 'International sustainable development policy:(Re) producing sexual norms through eco-discipline', *Gender, Place and Culture*, vol 21, no 8, pp. 1029–1044.

Foster, E.A. (2016) 'Eco-sexual normativity and queering ecologies', in G. Brown and K. Browne (eds.), *The Routledge Research Companion to Geographies of Sex and Sexualities*. Routledge, London.

Gaard, G. (ed.) (1993) *Ecofeminism: Women, Animals, Nature*. Temple University Press, Philadelphia, PA.

Gaard, G. (1997) 'Toward a queer ecofeminism', *Hypatia*, vol 12, no 1, pp. 114–137.

Gaard, G. (2015) 'Ecofeminism and climate change', *Women's Studies International Forum*, vol 49, pp. 20–33.

Garrard, G. (2010) 'How queer is green?', *Configurations*, vol 18, no 1, pp. 73–96.

Gosine, A. (2010) 'Non-white reproduction and same-sex eroticism: Queer acts against nature', in C. Mortimer-Sandilands and B. Erickson (eds.), *Queer Ecologies: Sex, Nature, Politics, Desire*. Kindle, Indiana University Press, Indianapolis, IN.

Greenpeace (2020) 'Rainbow warriors: How queer movements can inspire climate activists'. www.greenpeace.org.uk/news/rainbow-warriors-queer-movements-climate-activists/.

Chris, H. (2020) 'Climate catastrophe, animal agriculture, and the world's marginalised communities', in J. Feliz Brueck and Z. McNeill (eds.), *Queer and Trans Voices: Achieving Liberation through Consistent Anti-Oppression*. Sanctuary Publishers, Kindle.

Hamilton, C.L. (2019) *Veganism, Sex and Politics: Tales of Danger and Pleasure*. HammerOn Press, Bristol, UK.

Hays, C.M. (2019) 'The "Park" as racial practice: Constructing whiteness on Safari in Tanzania', *Environmental Values*, vol 28, no 2, pp. 141–170.

Hogan. K. (2010) 'Undoing nature: Coalition building as queer environmentalism', in C. Mortimer-Sandilands and B. Erickson (eds.), *Queer Ecologies: Sex, Nature, Politics, Desire*. Kindle, Indiana University Press, Indianapolis, IN.

Hatfield, J. and Dionne, J. (2014) 'Imagining ecofeminist communities via Queer Alliances in Disney's Maleficent', *Florida Communication Journal*, vol 42, no 2, pp. 81–98.

Inglehart, R.F. and Norris, P. (2016) 'Trump, Brexit, and the rise of populism: Economic have-nots and cultural backlash', *Harvard Kennedy School Faculty Research Working Paper*, Series RWP16-026.

Ko, A. and Ko, S. (2017) *Aphro-ism: Essays on Pop-Culture, Feminism, and Black Veganism from Two Sisters*. Lantern Books, Cheltenham.

Kotac, A. (2020) 'Is the Extinction Rebellion movement queer inclusive?', https:// meanshappy.com/the-extinction-rebellion-xr-movement-is-queer-inclusive/.

Merchant, C. (1980) *The Death of Nature*. Wildwood House, London.

Merchant, C. (2006) 'The scientific revolution and the death of nature', *Isis*, vol 97, no 3, pp. 513–533.

Morton, T. (2007) *Ecology without Nature: Rethinking Environmental Aesthetics*. Harvard University Press, Cambridge, MA.

Mortimer-Sandilands, C. (2005) 'Unnatural passions? A note toward a queer ecology', www. rochester.edu/in_visible_culture/Issue_9/sandilands.html, (accessed 13 August 2020).

Mortimer-Sandilands, C. and Erickson, B. (eds.) (2010) *Queer Ecologies: Sex, Nature, Politics, Desire*. Kindle, Indiana University Press, Indianapolis, IN.

Plumwood, V. (2002) *Feminism and the Mastery of Nature*. Routledge, London.

Oakley, A. (2016) 'Disturbing hegemonic discourse: Nonbinary gender and sexual orientation labeling on Tumblr', *Social Media+ Society*, vol 2, no 3, pp. 1–12.

Sandilands, C. (2002) 'Lesbian separatist communities and the experience of nature: Toward a queer ecology', *Organization & Environment*, vol 15, no 2, pp. 131–163.

Seagar, J. (1993) *Earth Follies: Coming to Feminist Terms with the Global Environmental Crisis*. Routledge, New York.

Sedgewick, E.K. (1993) *Epistemology of the Closet*. University of California Press, Berkeley, CA.

Simonsen, R.R. (2012) 'A queer vegan manifesto', *Journal for Critical Animal Studies*, vol 10, no 3, pp. 51–80.

Stanley, P. (2020) 'Unlikely hikers? Activism, Instagram, and the queer mobilities of fat hikers, women hiking alone, and hikers of colour', *Mobilities* vol 15, no 2, pp. 241–256.

The Green Party (2015), 'LGBTIQ Manifesto', www.greenparty.org.uk/resources/ LGBTIQ_Manifesto (accessed 02 August 2020)

Valentine, G. (2002) 'Queer bodies and the production of space', in D. Richardson and S. Seidman (eds.), *Handbook of Lesbian and Gay Studies*. Sage, London.

Warren, K.J. (1997) *Ecofeminism. Women, Culture Nature*. Indiana University, Bloomington, IN.

Whitworth, L. (2019) 'Goodbye Gauley Mountain, hello eco-camp: Queer environmentalism in the Anthropocene', *Feminist Theory*, vol 20, no 1, pp. 73–92.

7

THE DEMAND FOR RACIAL EQUALITY AND ENVIRONMENTAL JUSTICE

Learning from Bristol's Black and Green Programme

Roger Griffith and Gnisha Bevan

Introduction

'I can't breathe'. These were George Floyd's final tragic words after 8 minutes 46 seconds of torture. 'We can't breathe'. These are words that communities of colour have been stating for decades in the face of environmental racism (Ali, 2020). This chapter looks at how structural racism contributes both to environmental racism and exclusion from the 'mainstream' environmental movement. It draws on inspiring solutions from environmentalists of colour globally and, specifically, the Black and Green Programme based in Bristol, United Kingdom (UK). We conclude with recommendations on how to make 'mainstream' environmentalism more open to communities of colour.

There are a number of reasons why we should be able, and it might be preferable, to work in diverse groups. Firstly, growing evidence suggests that humans are likely to have evolved to be extremely empathetic and hardwired for cooperation (Karlberg, 2004; James, 2008; Rifkin, 2009; Unwin, 2018; Curry et al., 2019). Secondly, diverse teams and organisations make us more effective problem solvers (Phillips et al., 2006). Thirdly, we face complex, interconnected global problems that are impossible to solve without collaboration (Munasinghe, 2009). However, there are important systematic factors, including structural racism, that sometimes prevent us from working together to achieve our collective aims (Pederson, 2011).

When discussing race and identity, we acknowledge that there are no perfect terms. We accept valid concerns in categorising groups of people together (e.g. Bhopal, 2004). We have used several terms in this chapter. We use Black Asian and Minority Ethnic (BAME) in line with current UK policy language, whilst accepting its limitations (Okolosie et al., 2015; Banglawala, 2019). We also use the term 'Black' to refer to the sense of activism (Martin, 1991). We, additionally, embrace the terms 'communities of colour' and 'environmentalists of colour' as these are used more broadly across the globe as seen in 'Black Indigenous People of Colour' (BIPOC).

With regard to the term 'environmentalism', we refer to environmental activities in their broadest sense, recognising that this is contested (Gottlieb, 1993). We consider environmentalism to be a potentially inclusive term reflecting activities that increase our ability to live, work and play whilst protecting the earth and its creatures.

The mainstream environmental movement has been described as primarily white and middle-class (e.g. Agyeman, 2001; Alkon and McCullen, 2010; McLean, 2013). However, we question the idea that there is one form of 'mainstream' environmentalism and, instead, show that there are many movements all over the world. Therefore, we refer to 'mainstream' environmentalism in inverted commas to signify predominantly white and middle-class environmentalism.

Positionality

In this chapter, we draw on our experiences as BAME environmentalists. Roger has worked for three decades with underserved communities and as a race equality activist. He grew up in Lawrence Weston, Bristol, in a predominantly white, working-class area, a few miles from the smelting works and incinerators of Avonmouth. He can still see the smoke clouds belching into the sky and remembers playing football with an acrid taste in the air. Gnisha was raised in a middle-class environment and is of mixed heritage, with white working-class, and privileged Asian family members. She participated for a long time in 'mainstream' environmentalism, largely uncritically and in denial about structural racism, until returning home from international development work shocked her into exploring exclusion in 'mainstream' environmental circles.

Contributions of environmentalists of colour

Communities of colour are often perceived as not caring about the environment. However, evidence indicates that this does not reflect the reality. Communities of colour have contributed to environmentalism in innumerable ways: from nature connection to activism and from innovation to world-leading policies (Hawken, 2007; Senge et al., 2008). In the Global South, people on extremely low incomes care about, protect and maintain the environments they depend on (Martinez-Alier et al., 2016). Studies from the United States (US) indicate that communities of colour care more about the environment than white people because of living in poorer environmental conditions (Gibson-Wood and Wakefield, 2012; Elias et al., 2018). In the UK, more BAME students than white students have expressed a desire that their careers benefit the environment, even though they are the least likely to engage with 'mainstream' environmental organisations (NUS, 2018). At the local level, in Bristol, 89.2% of BAME people surveyed said that they care about climate change and had adopted environmentally friendly behaviours, such as using public transport and reducing energy and waste (BCC IPI, 2020) These reported behaviour changes were slightly higher than the Bristol average (*ibid.*). Yet, despite their environmental concerns, both in the US and the UK, barriers such as cost, time, elitism, irrelevance, lack of representation and absence of BAME leadership push the environment down the priority list for communities of colour (Paraskevopoulos et al., 2003; Pedersen, 2011; Griffith, 2016; NERC 2017).

In the UK, the 'mainstream' environmental sector is the least diverse, second only to agriculture (NUS, 2018), despite calls spanning almost half a century for this to change (e.g. Taylor, 1993). According to a recent study, just 3.1% of environmental professionals identified as BAME, a stark contrast to the 19.9% of BAME people that make up the UK workforce (NUS, 2018). Whilst 16% of employees in the UK statutory sector identify as BAME, only 3.85% identify as BAME at the Environment Agency and just 1.81% at Natural England (NUS, 2018).

This lack of BAME inclusion in these mainstream environmental policy-influencing and policy-making organisations is problematic because environmental priorities may vary according to people's backgrounds. The majority of UK BAME people can be considered as working-class (Cabinet Office, 2018), and therefore intersection with class issues should be considered (Gilroy, 2013) since middle-class and working-class people may have different concerns. Middle-class environmentalists may wish to protect *nature* and be 'engaged in bird watching, recycling, "buying green", hiking' (Allen et al. 2007, p. 127), whilst working-class environmentalists' primary focus could be the dangers that they encounter in their daily life (Griffith, 2016; NERC, 2017; Bell, 2020). These threats include a range of environmental issues: poor air quality, accident risk, waste contamination (e.g. proximity of incinerators and dumps), 'blighted' neighbourhoods, food poverty, chemical toxicity, fuel poverty, transport inequities, flooding and disaster, and inequalities in physical and mental health. Furthermore, there is a double injustice – that disadvantaged communities are least likely to create environmental harms, but most likely to suffer from them (Bell, 2020).

BAME communities have been using a holistic approach to combine environmentalism and social justice activities for several decades (Taylor 1993; Clarke and Agyeman, 2010; Agyeman, 2013). For example, living in harmony with the environment, laws of nature and the earth has been a central tenet of Rastafarianism from the 1930s (Nowakowski, 2016). In contrast, 'mainstream' environmentalism remains less holistic (Bell 2020; intersectionalenvironmentalist.com, 2020). Immigrant communities, such as Somali groups in the UK, have been found to bring their own household sustainability practices to the Global North and have knowledge that could make a positive contribution to green policy agendas (MacGregor et al., 2019). Globally, we find environmentalists of colour working in every sphere of environmentalism from activism to policy (Senge et al., 2008; Hawken, 2007; intersectionalenvironmentalist.com, 2020). This section explores some examples.

Many have heard of Greta Thunberg and her speeches in relation to the School Strikes for the Climate. However, there are other young environmental leaders of colour or from the Global South that we tend not to hear about, such as Vanessa Nakate (Uganda) (Evelyn, 2020), Xiye Bastida (Mexico) (Bastida, 2020), Ou

Hongyi (China) (Standaert, 2020), Isra Hirsi (USA), Licypriya Kangujam (India) (BBC, 2020b) and Jessica June Ahmed (UK) (Kale, 2020).

The first environmental activists were indigenous peoples resisting imperialism and colonisation (Figueroa Helland and Lindgren, 2016), struggling to maintain land rights, self-determine and live in relationship with past and future generations (MRG, 2019). Yet, tragically, despite the fact that indigenous peoples globally preserve 80% of all remaining plants and animals, and are fundamental to global sustainability (Gurria, 2017), they are the most likely to be killed for their activism (MRG, 2019). Today, the Atlas of Environmental Justice (EJAtlas, 2020) lists around 3,500 cases, many brought by indigenous movements, gaining important wins such as closing the Dakota Pipeline at Standing Rock (Boyle, 2020). In the US, environmental justice is intimately connected with civil rights struggles. For example, Black activists successfully shut down the Atlantic Pipeline in Virginia in July 2020 (Ortiz, 2020).

In terms of policy, several countries in the Global South have world leading (albeit controversial, in some cases) environmental policies, for example, Carbon Negative Bhutan (Youn, 2017), South Korea's Green Economy (UNEP, 2014), China's and India's renewables (UNFCC, 2017), and South Africa's Carbon Tax (Ntombela et al., 2019). Costa Rica prioritises environment over military, whilst Ecuador and Bolivia follow radical 'Vivir Bien/Buen vivir' approaches (Bell, 2017). Cuba, often left out from 'mainstream' sustainability debates, performs well on environmental justice (Bell, 2014) and has been found to be the world leader in sustainable development according to a new Sustainable Development Index (Hickel, 2018). Some of the highest numbers of women in parliament are found in Rwanda, Bolivia, Grenada and Namibia (Thornton, 2019), contributing to women's equality – a recommended solution for climate change (UNDP, 2016). Other examples of environmentalism abound, including traditional and cultural practices that have significantly less impact, in terms of ecological footprint, than modern practices (Vercoe and Brinkman, 2015; Hassan, 2020; Mukena, 2020).

This change and innovation offers much hope. But how much attention has this impactful work received? Often forgotten (Agyeman, 2001; Clarke and Agyeman, 2010; Alkon and McCullen, 2010), and under-reported in the mainstream media, it has even been erased. For example, Ugandan climate activist, Vanessa Nakate, was famously cropped out of pictures by journalists after posing with Greta Thunberg and other young, white, climate activists at a news conference at Davos (Evelyn, 2020).

Structural racism

Structural racism, also known as systematic racism and institutional racism, exists in wider society and, as such, is not left outside 'mainstream' environmentalism. Consisting of policies, beliefs, values, systems and procedures, structural racism comes from a power base that negatively affects communities of colour across every strata of their lives (Gilroy, 2013; Bhopal and Alibhai-Brown, 2018; Akala,

2018, Eddo-Lodge, 2018) and spans many countries of the globe (DiAngelo, 2016; 2019). Here we consider examples from the UK and the US in more detail before studying the wider global and historical contexts.

In the UK, education curriculums that do not reflect BAME experiences result in low grades, and attainment gaps marginalise communities of colour (Lammy, 2017). That education gap can continue disadvantage into the workplace with lower incomes for similar work, and harder to access capital for BAME businesses (Runnymede, 2020). Across social policy, similar patterns can be viewed (Bhopal and Alibhai-Brown, 2018), as seen in the Scarman Report (1982), the MacPherson Report (Lea, 2002) into the murder of Stephen Lawrence and the Runnymede Trust's 'Colour of Money' report (2020).

We also see that both in the UK and the US, police brutality towards Black and Brown people results in death and criminal injustice, including mass incarceration and disproportionate sentencing of Black and Brown people. This has rightly caught the headlines (Roberts and McVeigh, 2013; Koram, 2020; Stubley, 2020).

At the same time, many people of colour also die or fall ill due to environmental racism (Villarosa, 2020; Johnson, 2020b). In the US, the life expectancies of communities of colour have been shortened due to toxic air and water. The environmental justice literature is full of hundreds of examples, including the classic early work of Bullard (1990) and the United Church of Christ (1987). The most famous recent example in the US is that of Flint, Michigan (Ranganathan, 2016). The Flint community is 56.6% African-American, with 41.6% living in poverty (Pulido, 2016). The Michigan Civil Rights Commission (2017) reported that policymakers, government leaders and decision-makers at many levels failed the residents of Flint. The report states that the response was the result of implicit bias and the history of systemic racism (MCRC, 2017).

In another example from the US in 2005, Hurricane Katrina ripped through the southern Gulf states causing devastation not seen in decades (Hawkins and Maurer, 2009). On Roger Griffith's many visits to the area, to first document and then volunteer for an organisation in New Orleans (lowernine.org), working with the local community to rebuild their homes, he saw that the most devastated areas were low-income areas with large proportions of African-American home owners (lowernine.org, 2020). He has carried out rich cultural research in the State of Louisiana (Griffith, 2015). Plantation houses have been turned into museums that depict shocking atrocities of enslavement (Whitney Plantation, 2020). These plantation houses have been used in cinema, including for the Oscar winning film, 12 Years a Slave (2014). Over centuries, however, little has changed in terms of Black lives being valued. Centuries of systematic racism have been put in place from slavery, slave codes, Jim Crow and local bylaws on home ownership known as 'redlining' (Griffith, 2015). These lead to further systematic racism.

We can find similar parallels in environmental inequalities, such as those documented with regard to Cancer Alley between New Orleans and the state capital, Baton Rouge. Along Cancer Alley, an overabundance of environmental pollutants (Berry, 2003), such as petrochemicals, asbestos, coal fire and incinerator emissions, air pollution from traffic, housing containing hazardous lead, and cheap chemically contaminated land, can be found (Villarosa, 2020). Communities of colour are also living in cheaper housing that can more easily be swept away (Villarosa, 2020). And one parish in Cancer Alley has the highest death rate per capita in the US from coronavirus (Covid-19) (Harris, 2020). All these ongoing environmental and health problems contribute to the perception that communities of colours' lives do not matter.

The roots of structural racism are historical and economic, with enslavement, imperialism and colonialism forming the foundations of modern capitalism (Patel and Moore, 2018). In addition to divisions based on class, gender, etc., these systems needed racism in order to justify the oppression of millions of people (Akala, 2018; Haney Lopez et al., 2020). Capitalism changed relationships, creating divisions between people and between nature and 'society' (Brand and Wissen, 2012; Patel and Moore, 2018). Capitalist systems are now widely scrutinised for how they degrade the environment despite benefits of 'modernisation' (Klein, 2014), but discussions of the racism that allows capitalism to function are often missing (Haney Lopez et al., 2020). Enslavement, colonisation and imperialism introduced resource overexploitation, plantation monoculture, and ensuing bio-diversity, pollution and land degradation problems (Ax, 2011; Hornborg, 2016; Patel and Moore, 2018).

Yet, a vital part of this story is not simply environmental destruction, but the violation and destruction of people (Griffith, 2015; Akala, 2018; Eddo-lodge, 2018). For example, in 16th-century Madeira, where the entire island was deforested within 80 years, enslaved people and indentured labourers were violently abused in order to over-exploit the land (Patel and Moore, 2018). In Rapa Nui, it is likely that the capture and enslavement of local people contributed to environmental demise (Peiser, 2005). Deforestation in the Amazon continues to threaten the existence of indigenous people living there (Spring, 2019). Many of these systems continue uninterrupted, simply shape-shifting into neo-colonialism (Banerjee, 2003; Figueroa Helland and Lindgren, 2016). Racist constructions of the Global South allow the Global North to ignore the siting of extractive, polluting and destructive industries, exploitative labour practices, land grabbing and the extermination of activists (Banerjee, 2003; Haney Lopez et al., 2020).

Aside from more visible environmental racism, there are other reasons for the exclusion of communities of colour from 'mainstream' environmentalism. Whilst many 'mainstream' environmental groups may express a desire to work together with communities of colour, believing their doors to be open to everyone, this can fail in practice (BEN, 1994; BEN, 1997; Gibson-Wood and Wakefield, 2012; Bawden, 2015; Maclean, 2013; Pancost, 2016). As we shall see in the following sections, many of the behaviours excluding communities of colour are so ingrained that they happen unconsciously (Akala, 2018; Eddo-Lodge, 2018). In the UK, denial and confusion around racist structures is still prevalent (Akala, 2018; Eddo-Lodge, 2018) despite recent protests, such as Black Lives Matter and the inequalities highlighted by Covid-19, with a higher number of deaths per case for most BAME communities (BBC 2020a; Khan 2020; Bentley, 2020).

For BAME communities, the word 'environmentalism' conjures up images of middle-class people and activists whose concerns are far removed from their own daily lives (Griffith, 2016; Ujima Focus Group, 2018). For example, in Bristol, communities of colour perceived that 'saving the polar bears' in 'mainstream' environmentalism seemed to take precedence over the peace of mind derived from meaningful employment (Griffith, 2016). Solutions posed by privileged environmentalists may have little grounding in the daily realities of people from disadvantaged communities (BEN, 1994), particularly in the wake of Covid-19 (Griffith, 2020a). Furthermore, the language for discussing environmental issues is often complex and inaccessible (Brulle, 2010). Environmental information must be communicated effectively. It should be accessible, free of unnecessary jargon, and translated into community languages (Bevan, 2020a). Culturally appropriate language needs to be used that avoids 'othering' people. Terms like 'invasive species' or 'exotic' crops can make people uncomfortable and lead to exclusion (Peretti, 1998).

Our experience, corroborated by stories from environmentalists of colour in the Black Seeds Network (our support network for BAME environmentalists), has often thrown up 'mainstream' environmental worldviews that we consider problematic. Examples include a focus on 'fighting for a better world for our children', completely ignoring communities of colour affected now by climate change (Mitchell and Chaudhury, 2020; Mitchell et al., 2015); blaming communities of colour for a range of issues from migration (Hoffman, 2020) to overpopulation (Monbiot, 2020), with little discussion of these being fuelled by historical injustice or climate change; and a lack of awareness that, if not done appropriately, sustainable development can mean sustaining unjust economic systems and power relations (Banerjee, 2003; Mitchell and Chaudhury, 2020). It could be argued that these worldviews are the legacy of colonial superiority (McLean, 2013; Mitchell and Chaudhury, 2020).

Climate reparations, seeking to compensate those who suffer the worst effects of climate change yet did the least to cause it (Burkett, 2009), are not central to 'mainstream' environmental discourses. Neither are discussions regarding healing from past traumas stemming from the violence of oppression (Comas-Díaz et al., 2019). For many communities of colour, racial justice is non-negotiable; it cannot be abandoned to join an environmental movement that reinforces white superiority (Haney Lopez et al., 2020). Communities of colour, unsurprisingly, are reluctant to work with people that oppress them (Renger and Rees, 2016; Eddo-Lodge, 2018; Comas-Díaz et al., 2019; Bevan, 2020b). This lack of consideration of human needs and rights is at odds with 'mainstream' environmentalists' global perspective when it comes to thinking about ecology, climate and pollution (BEN, 1994; Agyeman, 2008).

Dealing with daily racism creates emotional, mental and physical labour and stress for communities of colour (Eddo-Lodge, 2018; Oluo, 2019; Johnson 2020a). We are often asked to educate on race equality issues but, when we speak about our experience, we face denial and backlash. When people benefit from privilege, it can often come as a shock and a threat when that privilege is exposed (Bhopal and Alibhai-Brown, 2018; Eddo-Lodge, 2018). Refusal to provide explanations on race equality issues can lead to being 'damned if we don't and damned if we do'. We may be asked to represent whole racialised communities or be experts on 'diversity and inclusion' (Rasool and Ahmed, 2020). All this work is usually unpaid, and takes our time away from rest and other important projects (Eddo-Lodge, 2018; Oluo, 2019; Griffith, 2020b; Johnson, 2020a).

We are not heard! – The ultimate micro-aggression

Representation can inspire, break down stereotypes and foster community cohesion (Rasool and Ahmed, 2020). Without it, key decisions are taken that adversely impact the health, wealth and well-being of disadvantaged communities. But structures embedded into capitalism misrepresent and erase communities of colour (Fraser 1995). These include symbolic, cultural and social patterns such as being misinterpreted and misrepresented by dominant cultures; being unrecognised and rendered invisible; and being stereotyped and disrespected in public and everyday life (Fraser, 1995; Gibson-Wood and Wakefield, 2012). Yosso (2005) describes a 'deficit mentality' as 'one of the most prevalent forms of contemporary racism' (p. 75), where knowledge, skills, abilities, aspirations, cultural wealth and communal bonds go untapped and unrecognised. Calls to remodel capitalism to achieve environmental aims often focus on economics, but we must also consider recognition justice (Fraser, 1995; Agyeman and Erickson, 2012; Agyeman, 2013). To combat this lack of representation, we must transform social norms by developing communities' strengths and cultural wealth, and contesting the deficit narrative (Yosso, 2005; Agyeman, 2013) through media representation and combatting stereotypes (Murji, 2006). We must connect 'mainstream' environmentalism with existing community activities and acknowledge the contributions of communities of colour (Agyeman, 2008; Clarke and Agyeman, 2010; Gibson-Wood and Wakefield, 2012).

There is always a risk that diversity and inclusion become tokenistic and do not address structural inequalities (Gibson-Wood and Wakefield, 2012; Griffith, 2016; Common Cause, 2018). A recurring complaint for environmentalists of colour is that, where we are included, we are listened to but not understood (Gibson-Wood and Wakefield, 2012; Griffith, 2016), leading to endless cycles of consultation but no real change. There is a 'need to ensure we are not having these same conversations in years to come. Break the cycle!' (Griffith, 2016, p. 6).

Many social and environmental projects lack time and money, and often require volunteers to work for free. Working without pay and lack of progression opportunities in the environmental sector can be off-putting for BAME people (NUS, 2018). Where BAME organisations set up independently, they may be the last to receive funding (Bundred, 2016; Bevan, 2019b). Responsibilities such as childcare can take precedence, so there is a lack of free time to volunteer (Griffith, 2016; Ketibuah-Foley and McKenzie, 2018). Moreover, working-class BAME people, particularly women, are more likely to have insecure and lower paid jobs (Cabinet Office, 2018). Currently, BAME people are more likely to have been adversely affected by Covid-19 (Griffith, 2020a). All of this leaves less time. Denial of these pressures and a lack of understanding can exacerbate the situation. For example, when Gnisha Bevan recommended paying BAME people for their time whilst attending a meeting of an environmental organisation, she was told '[w]ell the rest of us are giving our time for free'. This implied a lack of empathy and understanding of the structural inequalities that would prevent BAME people from participating.

The factors and barriers above coalesce into inappropriate solutions which do not meet the needs of BAME communities. We are asked to solve problems without resources and to engage in ways that are not on our terms and not in line with our priorities (BEN, 1994; Griffith, 2016). Without BAME leadership and decision-making, mainstream environmental organisations set their own priorities, not those of communities of colour (NERC, 2017). The next section discusses how a group of BAME environmentalists in Bristol have addressed the problems outlined above.

Ujima's Black and Green Project

In Bristol, one of the most affluent cities in the UK, there are high levels of multiple deprivation in BAME and working-class communities (BCCI, PI, 2020). This is reflected in participation gaps in economic and civic leadership, and inequalities in social and business settings affected by race, income and religion (CoDE, 2017). It is also true of environmental inequality, with some of the highest levels of pollution in the Lawrence Hill ward (Cameron, 2019), an area with a high proportion of BAME residents (Cameron, 2019; BCC IPI, 2020). In Bristol, generally, the highest levels of air pollution correlate to the areas with the highest BAME populations (Laville, 2019).

For a decade, Ujima Radio has delivered a number of projects to engage under-served communities and to challenge perceived thinking on environmentalism: from creating green citizen journalists, to a year of activity within the Bristol European Green Capital 2015. The project began in 2011 when Roger wrote a scoping paper for the Green Bristol Mayoral candidate highlighting some challenges in engaging communities of colour in environmentalism. This came to the attention of the Bristol Green Capital Partnership Director who became a key advocate. Together with fellow Ujima Director, Paul Hassan, an initiative began to champion and highlight those communities that felt they have no stake or status in the environmental movement.

In order to inspire tangible engagement with marginalised groups and challenge perceptions on what is 'Green' within Black communities, individuals, groups and organisations that were undertaking environmental projects were highlighted. Several community members aged between 18 and 30 years were trained in citizen journalism so as to report about their community and, thereby, redress the imbalance of stories on environmentalists of colour. They highlighted environmental injustices in their community, such as air quality along the M32 motorway, one of the most highly polluted corridors in Western Europe (Hart and Parkhurst, 2011). Marvin Rees, the current Mayor of Bristol, who grew up yards from the M32, has implemented a Clean Air Strategy stating a moral, ecological and legal duty to clean up the air (Laville, 2019).

In 2015, Ujima was unsuccessful in its application for a major European Green Capital funding grant. We used this as a catalyst to hold the city leaders accountable in terms of participation, diversity, inclusion, engagement and communication. Programme Manager, Julz Davis, launched the 'Green and Black' initiative (now known as Black and Green). Ujima worked for key questions about diversity, inclusion and communities to be included in the 2015 European Green Capital activities. The Black and Green (B&G) Ambassadors project built on this through two excellent leaders: Jasmine Ketibuah Foley and Zakiya Mackenzie. They delivered real change in a city where disadvantage and social injustice have been ingrained for years, as described below.

The philosophy underpinning the B&G Ambassadors Programme is to challenge, disrupt and support existing institutions to change their behaviour; to address unconscious biases; to discard practices that are implicitly exclusive; and to proactively think about how to create structures that are more inclusive (Ketibuah-Foley and Mckenzie, 2018). Importantly, following community feedback, the ambassadors were paid, trained and supported. A monthly radio programme was available digitally to engage environmental organisations and BAME organisations. In 2015, a radio debate broadcast to more than 22,000 listeners drew out reasons for exclusion and the barriers people encountered as well as existing community activities (Common Cause, 2018).

During their two-year pilot phase, the B&G Ambassadors created a body of work demonstrating the need and strong support for such a programme (Ketibuah-Foley and Mckenzie, 2018; Common Cause, 2018). It has been celebrated by city leaders including Bristol's elected mayors and community members alike (Ketibuah-Foley and Mckenzie, 2018). The B&G Ambassadors tapped into a reservoir of need and potential outlined in the next section.

BAME people in Bristol described 'walls' where the barriers between people are bigger than elsewhere and a culture of 'fear' where people in power maintain the status quo and shy away from addressing equality issues. They said there was not the space to talk about race and inequality within the green agenda (Griffith, 2016; Ketibuah-Foley and McKenzie, 2018). Many meetings were held in expensive, 'ethical' venues in the city centre, including an upmarket organic restaurant (Griffith, 2016). People from underserved communities found these locations inaccessible. Some BAME people can also feel there are 'no-go' areas for them in Bristol (Bevan, 2019a). Black and Green highlighted the importance of going to community spaces and organisations to meet people.

B&G Ambassadors partnered with community activists who had been voicing citizen concerns for decades, promoting girls education, or connecting inner city youth with nature (Mackenzie, 2020). They discovered that some well-known race activists, such as Roy Hackett, one of the leaders of Bristol Bus Boycott, and Peaches Golding, High Sheriff for the Queen, came from rural backgrounds. They were also passionate about the environment, but had never been asked about it (Hackett and Golding, 2016). A key finding was that BAME communities' concerns were much more related to their daily lives, for example, litter, or park maintenance, mirroring the concerns voiced by working-class people in recent UK research (Bell, 2020). Language was also important. Once BAME people understood unfamiliar terms, many realised they already took positive environmental actions. They enjoyed visiting green spaces in Bristol. Parks were a place to clear minds (Ujima Focus Group, 2018). However, poor upkeep and difficult access could put people off going to them (Ujima Focus Group, 2018). Nationally, 42% of BAME communities live in the most green space deprived neighbourhoods, with an average of just 9 m^2 of public green space (the size of an average garden shed) available. Children from the most deprived areas are 20% less likely to spend time outside than children in affluent areas (FOE, 2020). In the UK, many BAME communities lack access to the countryside (FOE 2020), but enjoy visiting through organised trips (BEN, n.d.). Bristol youths from BAME backgrounds may not be used to visiting nature, but do enjoy it when supported (Bevan, 2020b). On a trip to a local nature reserve, the B&G programme showed that BAME people find visitor attractions linked to their heritages and cultures appealing (Ketibuah-Foley and McKenzie, 2018). In addition, for BAME people, it is not always as relaxing as it should be to visit outdoor spaces. Collier (2019) argues that racism towards the first generations arriving in Britain after the Second World War (WW2) meant that people took refuge in cities and connection to the land was broken in one generation.

The B&G Ambassadors also carried out research with the African Caribbean community in inner-city Bristol and relevant environmental organisations (Common Cause, 2018). The B&G Ambassadors' research revealed environmental solutions that related to people's lives, like reducing energy costs through community renewables (B&G Ambassador, 2020); highlighting local 'ethical' businesses; discussing the merits of cycling (Mackenzie, 2020); and uncovering stories of Windrush generation elders growing food in allotments (B&G Ambassador, 2020). Delving into people's relationship with food, the Ambassadors found 'planet-friendly' diets, such as intersectional veganism, or Ital foods (Mackenzie, 2020). Academics working in environmental justice, and investigating links between the Sustainable Development Goals (SDGs) and the International Decade for People of African Descent appeared on their radio shows (B&G Ambassador, 2020).

The pilot project also sought to directly impact those decisions made by current leaders, mainly white, male, middle-class and university educated. They attended partner team meetings and spoke at major events and conferences including at the Cabot Institute, University of Bristol and the Bristol Green Capital Partnership. They also engaged with environmental leaders, including the then Mayor of Bristol, George Ferguson, his successor Marvin Rees, European Member of Parliament, Clare Moody, Professor Rich Pancost and US Presidential Candidate, Bernie Sanders (Griffith, 2016; Ketibuah-Foley and McKenzie, 2018).

The B&G Ambassadors were mentored and actively used reverse mentoring. They learnt about organisational structures tacitly available to the 'mainstream'. These outcomes shaped the decisions, campaigns and issues raised and, ultimately, where resources were allocated (Ketibuah-Foley and McKenzie, 2018).

The initial work of the B&G Ambassadors indicates that cultural transformation in Bristol's 'mainstream' environmental organisations will require their long-term committed joint engagement. Central to this has been a partnership approach that is rooted in the community to deliver change. The Black Seeds Network has subsequently been developed to act in symbiosis with the B&G Ambassadors Programme. By providing network, training, support, social media support and mentoring we are creating a new set of leaders and will be able to reach new stakeholders (Bevan, 2020b). The activists cited in this chapter during over a decade of activism have created a more inclusive participation and inclusion model that will help tackle gaps in economic and civic leadership and can be modelled nationally and internationally. These programmes continue to deliver meaningful actions that engage communities and challenge perceived thinking.

Conclusions and recommendations

This chapter highlights that partnership working rooted in the community is central to future progress. We have created better participation and inclusion models that will help tackle gaps in economic and civic leadership, and that can be modelled nationally and internationally. Based on the literature reviewed in this chapter, and our experience of involvement with the Black and Green Programme, we recommend the following:

Work together with BAME people to make decisions and write policies. Recruit and develop BAME leaders

Large organisations may express a desire for BAME leaders and talent but must take action on this (McGregor-Smith, n.d.). A lack of diversity in decision-making leads to inappropriate environmental solutions (Griffith, 2016). Positive action is required, with dedicated resources, expertise and time for capacity building, especially in fundraising (Griffith, 2016).

Resource BAME environmental projects: time, money, expertise, networks and connections

A lack of resources impedes the work of environmentalists of colour. This, plus the extra work of dealing with racism, adds burdens that can be difficult to surmount (Comas-Díaz et al., 2019). Well-resourced projects can make a huge contribution and are more sustainable (Griffith, 2016; Ketibuah-Foley and McKenzie, 2018).

Hold 'difficult conversations'

The Black and Green Programme found several ways to hold conversations about race and environmentalism, engaging in City Conversations, local radio, social media, community artists, forums, networks, and consulting local communities and businesses. Their leadership model meant new Ambassadors speaking at different environmental events and organisations (Davis, 2015; Griffith, 2016; Ketibuah-Foley and McKenzie, 2018; Common Cause, 2018; Patti and Pancost, 2018).

Conduct better research with under-represented communities

The area of BAME participation in environmentalism is under researched in the UK. In order to have better information and to find ways forward, we need more research. We recommend partnering *with* communities themselves. Black and Green's use of non-traditional research methods was highly engaging and accessible to a wide range of people (Common Cause, 2018).

Monitor equalities information at events and in environmental organisations

Since we 'manage what we measure', the stark reality can often be presented in figures. Publishing these can help accountability and spur action.

Include storytelling to change dominant narratives and break down stereotypical thinking

Recognise that there is not one 'mainstream' environmental movement. There are thousands of strands of environmentalism. Recognise that communities of colour have many effective environmental solutions that we can learn from and must respect (BEN, 1997; Hawken, 2007). Use the media and arts to tell the stories that have not been told about environmentalists of colour. Roger's own self-exploration of his identity in writing his story led to him feeling valued as his voice was heard. Since then, he has been able to inspire and engage others, creating employment and enterprise.

Not everyone finds scientific language accessible, but local people can spread meaningful messages especially in community languages (Griffith, 2016). Partner with community media outlets and community artists to share information in accessible language (Ketibuah-Foley and McKenzie, 2018; Common Cause, 2018). Build bridges into communities by using the cultural calendar for religious events, festivals and carnivals (Griffith, 2016).

Develop holistic environmental policies that meet people's needs and tackle inequality

Communities of colour are interested in the environment, but meeting basic needs can come first. It is important to think about how projects can be designed with and for communities so that they fit their agendas and consider how these can be resourced (Griffith, 2016).

Challenge stereotypes by widening discussions in environmentalism

Move beyond the local where appropriate. Looking only at local actions could have dangerous impacts for communities of colour elsewhere. Appreciate that many communities of colour have strong connections to the Global South (Koff, 2016). These connections can allow better discussions of the relationship between the Global North and the Global South.

Develop anti-racist approaches within organisations and projects

There is a risk that diversity and inclusion can be tokenistic, not going beyond the surface (Fredette et al., 2016). Many environmentalists of colour are seeking deeper moves towards equity and justice (Bevan, 2020b). It's the difference between 'being a better host' and 'becoming better humans' (Bevan, 2019b).

Implement the policies that have been recommended for decades!

Move beyond expressing desires to be inclusive. It is important to follow up with action. There have been decades of good intentions (Taylor 1993, McGregor-Smith, n.d.) so now we need action (McGregor-Smith, n.d.).

Finally

We have described how structural and environmental racism impacts communities of colour. Systematic racism is not only linked to police brutality and criminal injustice. Dismantling systematic racism includes tackling environmental injustice which intersects with policy issues such as housing, transport, homelessness, education, health and poverty. Ending environmental racism leads to a better quality of life, including healthy air and cleaner water.

The 'mainstream' environmental movement still carries too many hallmarks of being a white, middle-class domain. The inequalities first highlighted in 2011 in Roger's original report to the then Green leader remain and have got worse. There is a strong narrative of existing and potential engagement across communities of colour that needs to be recognised and developed. There needs to be a different mix of leaders, knowledge and voices, a continuous stream of positive project activity and a new narrative. Ujima's Black and Green programme has disrupted the way environmentalism takes place in Bristol and provides inspiration for what could happen elsewhere. Creating determined and extended engagement and leadership mechanisms with communities of colour presents significant opportunities for collective global action. In our work, we see a new energy emerging. Striving for racial and socio-economic justice will be a challenging yet ultimately rewarding process for us all. Will you join us?

References

12 Years a Slave (2014) [DVD]. Directed by Steve McQueen. River Road Entertainment, UK.

Agyeman, J. (2001) 'Ethnic minorities in Britain: Short change, systematic indifference and sustainable development', *Journal of Environmental Policy and Planning*, vol 3, no 1, pp. 15–30.

Agyeman, J., (2008) Toward a 'just' sustainability?', Continuum, vol 22, no 6, pp751-756.

Agyeman, J. (2013) *Introducing Just Sustainabilities,* Zed Books, London

Agyeman, J. and Erickson, J. (2012) 'Culture, recognition, and the negotiation of difference', *Journal of Planning Education and Research*, vol 32, no 3, pp. 358–366.

Akala (2018) *Natives: Race and Class in the Ruins of Empire.* Two Roads, London.

Allen, K., Daro, V. and Holland, D.C. (2007) Becoming an Environmental Justice Activist. In Sandler, R., and Pezzulo, P. (eds.), Environmental Justice and Environmentalism; The Social Justice Challenge to the Environmental Movement. MA: MIT, Cambridge

Alkon, A. and McCullen, C. (2010) 'Whiteness and farmers markets: Performances, perpetuations …contestations?' *Antipode*, vol 43, no 4, pp. 937–959.

Ax, C.F. (2011) *Cultivating the Colonies: Colonial States and Their Environmental Legacies*, Ohio University Press, Athens, OH.

B&G Ambassador (2020) *Interview on the Successes of the Black and Green Programme.* 8 September, Bristol.

Banerjee, B. (2003) 'Who sustains whose development? Sustainable development and the reinvention of nature', *Organization Studies*, vol 24, no 1, pp. 143–180.

Bastida, X. (2020) 'My name is not Greta Thunberg: Why diverse voices matter in the climate movement', *The Elders*, 19 June.

Bawden, T. (2015) 'Green movement must escape its 'white, middle-class ghetto', says Friends of the Earth chief Craig Bennett', *The Independent*, 5 July.

BBC (2020a) 'Black lives matter: We need action on racism not more reports, says David Lammy', *BBC*, 15 June.

BBC (2020b) 'India climate activist Licypriya Kangujam on why she took a stand', *BBC*, 6 February.

BCC IPI (2020) Bristol City Council Insight, Performance and Intelligence 'Bristol Quality Of Life Survey 2019/20', Bristol City Council, Bristol.

Bell, K. (2014) *Achieving Environmental Justice: A Cross-National Analysis.* Policy Press, Bristol.

Bell, K. (2017) '"Living Well' as a path to social, ecological and economic sustainability', *Urban Planning*, vol 2, no 4, pp. 19–33.

Bell, K. (2020) *Working-Class Environmentalism: An Agenda for a Just and Fair Transition to Sustainability.* Palgrave Macmillan, London.

BEN (1994) 'Involving urban communities in the environment', http://ben-network.org.uk//uploaded_Files/Ben_1/ben_file_1_3.pdf.

BEN (1997) 'Building multi-culturalism as a framework for ethnic environmental participation', http://ben-network.org.uk//uploaded_Files/Ben_1/Ben_file_1_27.pdf.

BEN (n.d.) 'Access to the countryside trips – Report (excerpt)', http://ben-network.org.uk//uploaded_Files/Ben_1/Ben_file_1_26.pdf.

Bentley, G. (2020) 'Don't blame the BAME: Ethnic and structural inequalities in susceptibilities to COVID-19', *American Journal of Human Biology*, vol 32, no 5, pp. 1–5.

Berry, G. (2003) 'Organizing against multinational corporate power in Cancer Alley', *Organization and Environment*, vol 16, no 1, pp. 3–33.

Bevan, G. (2019a) *Notes on Bristol Equality Network meeting.* 26 September, Bristol.

Bevan, G. (2019b) *Notes on Race Equality in Nature: The Next Generation 13–30 Conference.* 2 October. Black2Nature, Bristol.

Bevan, G. (2020a). *Notes on the Climate & Our Community Conference*. Bristol.

Bevan, G. (2020b) *Black Seeds Network Meeting*. 15 June, Bristol.

Bhopal, K. and Alibhai-Brown, Y. (2018) *White Privilege*. Policy Press, Bristol.

Bhopal, R. (2004) 'Glossary of terms relating to ethnicity and race: For reflection and debate', *Journal of Epidemiology and Community Health*, vol 58, no 6, pp. 441–445.

Boyle, L., (2020) 'Standing Rock tribe celebrates 'significant win' over Trump in pipeline court ruling', Independent Online, 26 March

Brand, U. and Wissen, M. (2012) 'Global environmental politics and the imperial mode of living: Articulations of state–capital relations in the multiple crisis', *Globalizations*, vol 9, no 4, pp. 547–560.

Bristol Green Capital (2015) https://bristolgreencapital.org/who-we-are/european-green-capital-award/

Bruce-Lockhart, A. (2016) *World Economic Forum*, 29 July. www.wefcrum.org/agenda/2016/07/greenest-happiest-country-in-the-world/.

Brulle, R. (2010) 'From environmental campaigns to advancing the public dialog: Environmental communication for civic engagement', *Environmental Communication*, vol 4, no 1, pp. 82–98.

Bullard, R. (1990) *Dumping in Dixie: Race, Class and Environmental Quality*. Westview Press, Boulder, CO.

Bundred, S. (2016) 'Report to Bristol City Council: Review of Bristol 2015 European Green Capital year', Bristol.

Bunglawala, Z. (2019) Please, don't call me BAME or BME!. *Civil Service Blog*, https://civilservice.blog.gov.uk/2019/07/08/please-dont-call-me-bame-or-bme/.

Burkett, M. (2009), 'Climate reparations', *Melbourne Journal of International Law,* vol. 10, no. 2, pp. 509–542.

Cabinet Office (2018) 'Race Disparity Audit Summary findings from the Ethnicity Facts and Figures website October 2017', www.ethnicity-facts-figures.service.gov.uk/static/race-disparity-audit-summary-findings.pdf.

Cameron, A., (2019) 'Air pollution leads to five premature deaths a week in Bristol', Bristol Live, 19 November

Centre on Dynamics of Ethnicity (CoDE) in collaboration with the Runnymede Trust (2017) 'Bristol: A city divided? Ethnic Minority Disadvantage in Education and Employment', Centre on Dynamics of Ethnicity (CoDE) in collaboration with The Runnymede Trust, Manchester.

Clarke, L. and Agyeman, J. (2010). 'Is there more to environmental participation than meets the eye? Understanding agency, empowerment and disempowerment among black and minority ethnic communities', *Area*, vol 43, no 1, pp. 88–95.

Collier, B. (2019) 'Black absence in green spaces', *The Ecologist*, 10 October.

Comas-Díaz, L., Hall, G.N. and Neville, H.A. (2019) 'Racial trauma: Theory, research, and healing: Introduction to the special issue', *The American Psychologist,* vol 74, no 1, pp. 1–5.

Common Cause (2018) 'Green and Black – PhotoVoice: Through My Lens', A collaboration between Ujima Radio and University of Bristol.

Curry, O., Mullins, D. and Whitehouse, H. (2019) 'Is it good to cooperate? Testing the theory of morality-as-cooperation in 60 societies', *Current Anthropology* vol 60, no 1, pp. 47–69.

Davis, J. (2015) 'Origins', Bristol, www.ideasfestival.co.uk/wp-content/uploads/2016/11/Julz-Davis-Green-and-Black-Initiative.pdf.

DiAngelo, R. (2016) *What Does It Mean to be White?*, Peter Lang, New York.

DiAngelo, R.J. (2019) *White Fragility: Why it's so Hard for White People to Talk about Racism*, Allen Lane, UK.

Eddo-Lodge, R. (2018) *Why I'm No Longer Talking to White People about Race*, Bloomsbury, London.

EJAtlas (2020) 'Mapping environmental justice', https://ejatlas.org/.

Elias, T., Dahmen, N., Morrison, D., Morrison, D. and Morris, D. (2018). 'Understanding climate change perceptions and attitudes across racial/ethnic groups', *Howard Journal of Communications*, vol 30, no 1, pp. 38–56.

Evelyn, K. (2020) '"Like I wasn't there": Climate activist Vanessa Nakate on being erased from a movement', *The Guardian*, 29 January.

Figueroa Helland, L. and Lindgren, T. (2016) 'What goes around comes around: From the coloniality of power to the crisis of civilization', *Journal of World-Systems Research*, vol 22, no. 2, pp. 430–462.

Fraser, N. (1995), 'From redistribution to recognition? Dilemmas of justice in a 'Post-Socialist' age', *New Left Review*, no 212, p. 68.

Fredette, C., Bradshaw, P. and Krause, H. (2016) 'From diversity to inclusion: A multimethod study of diverse governing groups', *Nonprofit and Voluntary Sector Quarterly*, vol 45, no 1, pp. 28S–51S.

Friends of the Earth (2020) 'Access to green space in England. Are you missing out?', https://friendsoftheearth.uk/nature/access-green-space-england-are-you-missing-out.

Gibson-Wood, H. and Wakefield, S. (2012) '"Participation", white privilege and environmental justice: Understanding environmentalism among Hispanics in Toronto', *Antipode*, vol 45, no 3, pp. 641–662.

Gilroy, P. (2013) *There Ain't No Black In the Union Jack*. Routledge, London.

Gottlieb, R. (1993) 'Reconstructing environmentalism: Complex movements, diverse roots', *Environmental History Review*, vol 17, no 4, pp. 1–19.

Griffith, R. (2015) *My American Odyssey: From the Windrush to the Whitehouse*. Silverwood, Bristol.

Griffith, R. (2016) 'The Green and Black report: A report on Ujima Radio's initiative to involve Black Minority Ethnic communities in the Green agenda during Bristol European Green Capital 2015', Ujima Radio, Bristol.

Griffith, R. (2020a) 'How BAME groups have been disproportionately affected by Covid-19', *B24/7*, 22 May.

Griffith, R. (2020b). 'It is time to lift the rock of racism from our bodies', *Bristol Live*, 16 June.

Gurria, E. (2017) 'Celebrating indigenous peoples as nature's stewards', *UNDP Blog*, 2 May.

Hackett, R. and Golding, P. (2016) 'Soul Music Live – A change is gonna come', BBC Radio 4 and Ujima Radio, 29 November.

Haney Lopez, I., Elliott-Cooper, A. and Hallam, R. (2020) 'Building solidarity across race and class', Extinction Rebellion, www.youtube.com/watch?v=3U7PYZCObtk.

Harris, K. (2020) '"A terrible price": The deadly racial disparities of Covid-19 in America', *New York Times*, 29 April.

Hart, J. and Parkhurst, G. (2011) 'Driven to excess: Impacts of motor vehicles on the quality of life of residents of three streets in Bristol UK', *World Transport Policy and Practice*, vol 17, no 2, pp. 12–30.

Hassan, D., (2020) 'Who are the silent voices in the green movement?', *Rife Magazine*, 27 January

Hawken, P. (2007) *Blessed Unrest*, Penguin Books, New York.

Hawkins, R. and Maurer, K. (2009) 'Bonding, bridging and linking: Ow social capital operated in New Orleans following Hurricane Katrina', *British Journal of Social Work*, vol 40, no 6, pp. 1777–1793.

Hickel, J. (2018) 'Is it possible to achieve a good life for all within planetary boundaries?', *Third World Quarterly*, vol 40, no 1, pp. 18–35.

Hoffman, R. (2020) 'Climate change, migration and urbanisation: Patterns in sub-Saharan Africa', *The Conversation*, 2 November.

Hornborg, A. (2016) 'Artifacts have consequences, not agency', *European Journal of Social Theory*, vol 20, no 1, pp. 95–110.

Intersectionalenvironmentalist.com (2020) 'Meet Our Community + Topic Leaders', www.intersectionalenvironmentalist.com/community-topic-leaders.

James, O. (2008) *The Selfish Capitalist*. Random House, London.

Johnson, A. (2020a) 'I'm a black climate expert. Racism derails our efforts to save the planet', *Washington Post*, 3 June.

Johnson, T. (2020b) 'Kamala Harris understands how to fight systemic racism through environmental justice', *The Hill*, 17 August.

Kale, S. (2020) '"We need to be heard": The BAME climate activists who won't be ignored', *The Guardian*, 9 March.

Karlberg, M. (2004) *Beyond the Culture of Contest*. George Ronald, Oxford.

Ketibuah-Foley, J. and McKenzie, Z. (2018) 'Green & Black Ambassadors Pilot Project Report', Ujima Radio, Bristol Green Capital, Up Our Street, University of Bristol Cabot Institute, Bristol.

Khan, O. (2020) 'Coronavirus exposes how riddled Britain is with racial inequality', *The Guardian*, 20 April.

Klein, N. (2014) *This Changes Everything: Capitalism vs. the Climate*. Allen Lane, St Ives.

Koff, H. (2016) 'Diaspora philanthropy in the context of policy coherence for development: Implications for the post-2015 Sustainable Development agenda', *International Migration*, vol 55, no 1, pp. 5–19.

Koram, K. (2020) 'Systemic racism and police brutality are British problems too', *The Guardian*, 4 June.

Lammy, D. (2017) 'The Lammy Review', www.gov.uk/government/publications/lammy-review-final-report.

Laville, S. (2019) 'Air pollution kills five people in Bristol each week, study shows', *The Guardian*, 18 November.

Lea, J. (2002) 'The Macpherson report and the question of institutional racism', *The Howard Journal of Criminal Justice*, vol 39, no 3, pp. 219–233.

lowernine.org. (2020) 'Home', https://lowernine.org/.

Mackenzie, Z. (2020) 'Ujima Green and Black showreel – Zakiya Mecca', http://zakiyamckenzie.com/2017/11/ujima-green-and-black-showreel/.

Martin, B. (1991) 'From negro to black to African American: The power of names and naming', *Political Science Quarterly*, vol 106, no 1, p. 83.

Martinez-Alier, J., Temper, L., Del Bene, D. & Scheidel, A. (2016) 'Is there a global environmental justice movement?', *The Journal of Peasant Studies*, vol 43, no 3, pp. 731–755.

MacGregor, S., Walker, C. and Katz-Gerro, T. (2019) '"It's what I've always done": Continuity and change in the household sustainability practices of Somali immigrants in the UK', *Geoforum*, vol 107, pp. 143–153.

McGregor-Smith, R. (n.d.) 'The time for talking is over. Now is the time to act', UK, www.basw.co.uk/resources/time-talking-over-now-time-act-race-workplace.

McLean, S. (2013) 'The whiteness of green: Racialization and environmental education', *The Canadian Geographer/ Le Géographe Canadien*, vol 57, no 3, pp. 354–362.

Michigan Civil Rights Commission (2017) 'The Flint Water Crisis: Systemic racism through the lens of flint', Michigan Civil Rights Commission.

Minority Rights Group International (2019) 'Minority and Indigenous Trends 2019', MRGI, London.

Mitchell, A. and Chaudhury, A. (2020) 'Worlding beyond "the" "end" of "the world": White apocalyptic visions and BIPOC futurisms', *International Relations*, vol 34, no 3, pp. 309–332.

Mitchell, G., Norman, P. and Mullin, K. (2015) 'Who benefits from environmental policy? An environmental justice analysis of air quality change in Britain, 2001–2011', *Environmental Research Letters*, vol 10, no 10, pp. 1–19.

Monbiot, G. (2020) 'Population panic lets rich people off the hook for the climate crisis they are fuelling', *The Guardian*, 26 August.

Moser, S. and Dilling, L. (2004) 'Making climate HOT', *Environment: Science and Policy for Sustainable Development*, vol 46, no 10, pp. 32–46.

Mukena (2020) 'Sustainable practices acclaimed by environmental groups are what Black people "have been doing for years"', *Bristol Live*, 7 September.

Munasinghe, M. (2009) *Sustainable Development in Practice' Sustainomics Methodology and Applications*. Cambridge University Press, Cambridge.

Murji, K. (2006) 'Using racial stereotypes in anti-racist campaigns', *Ethnic and Racial Studies*, vol 29, no 2, pp. 260–280.

NERC (2017) 'Green and Black ambassadors: Tackling inequality in Bristol', *Planet Earth*.

Nowakowski, K. (2016) 'For Rastas, eating pure food from the earth is a sacred duty', *National Geographic*, 19 July. https://www.nationalgeographic.com/culture/food/the-plate/2016/07/for-rastas--eating-from-the-earth-is-a-sacred-duty/

Ntombela, S.M., Bohlmann, H.R. and Kalaba, M.W. (2019) 'Greening the South Africa's economy could benefit the food sector: Evidence from a Carbon Tax Policy Assessment', *Environmental and Resource Economics*, vol 74, no 2, pp. 891–910.

NUS, The Equality Trust, IEMA (2018) 'Race, inclusivity and environmental sustainability', London.

Okolosie, L., Harker, J., Green, L. and Dabiri, E. (2015) 'Is it time to ditch the term "black, Asian and minority ethnic" (BAME)?', *The Guardian*, 22 May.

Oluo, I. (2019) *So You Want To Talk about Race*. SEAL, New York.

Ortiz, E. (2020) 'Atlantic Coast Pipeline cancelled after years of delays, accusations of environmental injustice', *NBC News*, 6 July.

Pancost, R. (2016) 'Cities lead on climate change', *Nature Geoscience*, vol 9, pp. 264–266.

Paraskevopoulos, S., Korfiatis, K. and Pantis, J. (2003) 'Social exclusion as constraint for the development of environmentally friendly attitudes', *Society and Natural Resources*, vol 16, no 9, pp. 759–774.

Patel, R. and Moore, J. (2018) *A History of the World in Seven Cheap Things*. University of California Press, California, UC.

Patti, D. and Pancost, R. (2018) 'How engaging citizens can help to shape green cities', *Cabot Institute for the Environment Blog*, 3 July, http://cabot-institute.blogspot.com/2018/07/how-engaging-citizens-can-help-to-shape.html.

Pedersen, O. (2011) 'Environmental justice in the UK: Uncertainty, ambiguity and the law', *Legal Studies*, vol 31, no 2, pp. 279–304.

Peiser, B. (2005) 'From genocide to ecocide: The rape of Rapa Nui', *Energy and Environment*, vol 16, no 3–4, pp. 513–539.

Peretti, J.H. (1998) 'Nativism and nature: Rethinking biological invasion', *Environmental Values*, vol 7, no 2, pp. 183–192.

Phillips, K.W., Northcraft, G.B. and Neale, M.A. (2006) 'Surface-level diversity and decision-making in groups: When does deep-level similarity help?', *Group Processes & Intergroup Relations*, vol 9, no 4, pp. 467–482.

Pulido, L. (2016) 'Flint, environmental racism, and racial capitalism', *Capitalism Nature Socialism*, vol 27, no 3, pp. 1–16.

Ranganathan, M. (2016) 'Thinking with flint: Racial liberalism and the roots of an American water tragedy', *Capitalism Nature Socialism*, vol 27, no 3, pp. 17–33.

Rasool, Z. and Ahmed, Z. (2020) 'Power, bureaucracy and cultural racism', *Critical Social Policy*, vol 40, no 2, pp. 298–314.

Renger, D, and Reese, G, (2016) 'From equality-based respect to environmental activism: Antecedents and consequences of global identity', *Political Psychology*, Vol 38, no 5, pp.867-879.

Rifkin, J. (2009) *The Empathic Civilization*. Polity Press, Cambridge.

Roberts, D. and McVeigh (2013) 'Eric Holder unveils new reforms aimed at curbing US prison population', *The Guardian,* 12 August.

Runnymede Trust (2020) 'The Colour of Money How Racial Inequalities Obstruct a Fair and Resilient Economy', Runnymede Trust, London

Scarman, L.S. and Baron (1982) 'The Scarman report: The Brixton disorders, 10–12 April 1981', Penguin, Harmondsworth.

Senge, P., Smith, B., Kruschwitz, N., Laur, J. and Schley, S. (2008) *The Necessary Revolution*. Nicholas Brealey Publishing, London.

Spring, J. (2019) 'Brazil's indigenous people swear to fight for Amazon "to last drop of blood"', *Reuters*, 23 August.

Standaert, M. (2020) 'China's first climate striker warned: Give it up or you can't go back to school', *The Guardian*, 20 July.

Stubley (2020) 'Authorities must apologise for "knee on neck" police arrest, lawyer says'. *The Independent*, 18 July.

Taylor, D. (1993) 'Minority environmental activism in Britain: From Brixton to the Lake District', *Qualitative Sociology*, vol 16, no 3, pp. 263–295.

Thornton, A. (2019) 'These countries have the most women in parliament', *WEF,* 12 February.

Ujima Focus Group (2018) Interviewed by Ujima Radio. 'Photovoice "Through my lens" project', Unpublished, Bristol.

UNDP (2016) 'Gender and Climate Change: Overview of Linkages between Gender and Climate Change', UNDP, New York.

UNEP (2014) 'Korea's pathway to a green economy', UN Environmental Program, Geneva.

UNFCC (2017) 'China and India lead global renewable energy transition', UNFCCC, 21 April.

United Church of Christ Commission for Racial Justice (1987) 'Toxic wastes and race in the United States', www.ucc.org/.

Unwin, J. (2018) 'Kindness, emotions and human relationships: The blind spot in public policy', Carnegie UK Trust, Dunfermline.

Verooe, R. and Brinkman, R. (2015) 'Comparing sustainability in the Global North and South to uncover meaning for educators', *The Journal of Sustainability Education*, 19 March.

Villarosa (2020) 'Pollution is killing black Americans', *New York Times*, 28 July.

Whitney Plantation (2020) 'Whitney Plantation', www.whitneyplantation.org.

Yosso, T. (2005) 'Whose culture has capital? A critical race theory discussion of community cultural wealth', *Race Ethnicity and Education*, vol 8, no 1, pp. 69–91.

Youn, S. (2017) 'Visit the world's only carbon-negative country Bhutan has built sustainability into its national identity', *National Geographic,* www.nationalgeographic.com/travel/destinations/ashutantan/carbon-negative-country-sustainability/.

8

THE FORGOTTEN GENERATION

Older people and climate change

Gary Haq

Introduction

The climate emergency is one of the greatest challenges to humanity. Yet for many, the prospect of global climate chaos and the impact this will have on millions of people is often seen as a challenge for tomorrow. Although environment plays a role in everyone's life, the extent to which environmental pollution affects a particular individual will be influenced by socio-economic status, gender, race, geography and age. This is especially the case for climate change.

Global mean temperature is likely to reach 1.5°C above pre-industrial levels between 2030 and 2052 if it continues to increase at the current rate. Rapid, far-reaching and unprecedented changes in all aspects of society are needed to keep below 1.5°C heating (IPPC, 2018). As the planet warms, we can expect increasing climate variability that will have direct and indirect effects on human health and well-being (Bouzid et al., 2013). Warmer global temperatures are likely to increase the frequency, severity and duration of extreme weather events, such as heat waves, tropical cyclones, flooding, sea level rise and severe storms (IPPC, 2014).

This changing climate is occurring at a time when the global population is getting older. In 2050, one in six (16%) people in the world will be over the age of 65 years, while the number of persons aged 80 years or over is projected to almost triple, from 143 million in 2019 to 426 million in 2050 (UNDESA, 2019). Older people face a higher risk of climate impacts compared to the rest of the population due to some older people having issues related to mobility, disability and ill-health (Harper, 2019). The surge in the population of post-war baby boomers – adults born between 1946 and 1964 (Sands, 2018) – is having a particular impact on society not only because of their numbers, but also with regard to the different values and attitudes they hold. Some are bringing higher levels of consumption to middle and later life. They are re-inventing old age, basing it on new consumption and leisure-orientated lifestyles, where travel and cosmopolitanism are key features

(Biggs et al., 2007). These new values and related actions have consequences for greenhouse gas (GHG) emissions.

Global youth-led climate strikes, lawsuits demanding government action to reduce GHG emissions and advocacy by young climate activists, such as Greta Thunberg, show that young people are concerned about their future (Wu et al., 2020). Thunberg has gained international recognition with her call for action to address climate change. Ahead of the 2020 World Economic Forum in Davos, Thunberg, together with other young climate activists, wrote an op-ed which said '[y]oung people are being let down by older generations and those in power' in reference to the lack of action to combat climate change (Thunberg, 2020, n.p.). Some climate activists have been accused of adopting an intergenerational unfairness narrative, blaming older people for overconsumption and climate inaction (Curzon, 2020; Karpf, 2020). Other youth figures, such as singer-song writer Billie Eilish, have spurred on the movement by saying '[h]opefully the adults and the old people start listening to us [about climate change]… Old people are gonna die, and don't really care if we die, but we don't wanna die yet' (Trendell, 2019, np).

Ageism and the climate movement

Stereotypes of older people tend to be rooted in ageism, which is different from some other forms of prejudice in that ageism represents bias and discrimination by members of one group against the members of a second group which the first group will one day join (Donizetti, 2019). Self-esteem and ageing anxiety have been identified as significant predictors of negative stereotyping of older people (Tasdemir, 2020). Young adults, anxious about their future, attribute to older people the negative stereotypes that they fear will describe their own futures (Allan et al., 2009).

The coronavirus (COVID-19) pandemic may have reinforced the ageist narrative. During the pandemic, older people were portrayed in the media as a homogeneous group associated with death, dependency and high vulnerability. This could further justify discriminatory practices against the older generation (Bravo-Segal and Villar, 2020).

While older people are vulnerable to climate impacts, it is necessary to overcome stereotypes of being incapable, passive or disinterested (Haq et al., 2010). Policy discussions at an international and local level are reinforcing this negative view of older people by failing to incorporate them into policy debates. The UN Sustainable Development Goals set out targets to 'promote prosperity while protecting the planet', but only mention older people twice (Curzon, 2020). This chapter will address the issue of older people and climate change and whether climate concern and action is just for the young.

Climate change and older people

Virtually every country in the world is experiencing growth in the number and proportion of older people in their population (UN, 2019). An ageing population is a social transformation that will have implications for housing, travel, health and social care. Older people represent a growing share of the consuming and voting public. Their lifestyle choices have an impact and they vote in high numbers, so their attitude to climate change matters (Frumkin et al., 2012).

A changing climate is expected to have adverse effects on natural and human systems. The risk and harm resulting from climate change will not be evenly distributed (Harper, 2019). Although many older people are healthy and socially and economically active, others are not. Some may be physically, financially and emotionally less able to deal with the effects of a changing climate compared with the rest of the population (Haq et al., 2008; Sánchez-Gonzalez and Chávez-Alvardo, 2016). It is also important to recognise the experience, knowledge and potential of older generations and the role they could play in climate action.

Understanding the factors that contribute to older people's vulnerability and resilience can strengthen the capacity of governments to prevent and minimise climate-related impacts on this demographic group (Brenan et al., 2019; Doyle et al., 2019; Baldwin et al., 2020; Malak et al., 2020). This requires not only being aware of potential increased vulnerabilities of older adults to a changing climate, but also acknowledging the environmental impact of their lifestyle choices and recognising their contribution to the climate movement. Older people can therefore be seen not only as casualties of climate change, but also as champions of the response to this challenge.

Carbon culprits

When 16-year-old Greta Thunberg addressed the 2019 UN Climate Action Summit in New York claiming, '[y]ou have stolen my dreams and my childhood …' (NPR, 2019). Her words drew a line between 'you', the older generations, and 'us', the young people. The implication of this was that the older generations are to blame for climate change and the young and future generations will have to live with the consequences (Unruh, 2019). Moody (2017) argues that older people bear the responsibility for climate change because they have contributed to it more than any other age groups. Those who are old today have lived lives in which economic abundance has been based on carbon pollution and they should, therefore, bear disproportionate responsibility for the problem.

The older generation is not a homogeneous group for each older person has particular characteristics reflective of their life stage. Haq et al. (2007) categorised the United Kingdom (UK) older adults into three general groups: Baby boomers (aged 50–64 years), seniors (aged 65–74 years) and elders (aged 75+ years). Baby boomers are the first generation to grow up in the consumer society and many are in their highest earning years. They have different attitudes to ageing compared to their predecessors. In contrast, seniors have lived through the Second World War and grew up in years of austerity. The majority have a low income and tend to be prompt bill payers, debt averse and dislike waste. This group has experienced major lifestyle and life stage changes such as retirement and bereavement. The elders share many of the characteristics of the seniors. They have lived through great hardships during the 20th century. They are aware of their own mortality and the concept of death. They may be coping with increasing care needs and declining health.

Housing, food, energy and personal travel all impact energy use and carbon emissions. An analysis of UK age-related household consumption (Haq et al. 2007) found that baby boomers had the largest carbon footprint, with car dependency, holidays abroad and eating out as key carbon-intensive activities, while home energy use was a major contributor to the carbon footprint of older adults (aged 75+ years) (Hamza and Gilroy, 2011). A study of intergenerational emissions found that French baby boomers were not only better off than other generations but consumed more and lived in energy-inefficient homes (Chancel, 2014).

Age-specific estimates of per capita GHG emissions in the United States (US) for a set of selected carbon-intensive goods (e.g. electricity, natural gas, transport, food and clothes) also showed that average emissions increase with age (Estiri and Zagheni, 2019). This continues until people reach their late 60s, after which per capita emissions decrease, with use of energy-intensive goods decreasing in later life (Zagheni, 2011).

Poor health is often associated with old age, which can result in lower use of environmentally friendly transport (e.g. walking and cycling) as well as energy-intensive modes such as driving and flying (Okada, 2012). Furthermore, domestic electricity consumption increases (Romanach et al. 2017; Buchs et al., 2018). With more older people working in later life, there is the potential to reduce the negative effect of ageing on the economy, yet this may not be the case for GHG emissions. Some studies suggest that early retirement can lower energy use and GHG emissions from the labour sector (Wei et al., 2018). However, other studies imply that, if retirement is postponed, emission-reducing effects in later life could be reduced (O'Neil et al., 2012). An individual's pattern of consumption changes over their life course, reflecting wealth, age, health, social needs and geographical location (Venn et al., 2017; Hitchings et al., 2015, 2018). Not every older person's consumption pattern is the same. The claim of intergenerational inequality may therefore present a false picture of all older adults being rich and lucky, and millennials being downtrodden and poor.

We need to understand better how different stages of ageing impact energy use and carbon emissions. Berstow (cited in Willets and Berstow, 2019) argues that blaming older people encourages a sentiment of grievance and fatalism among young people, incited to resent their elders rather than engaging them in the struggle to shape the future (Willets and Berstow, 2019, n.p.). She concludes:

> Whether Baby Boomers or Millennials, people do not engage with the world as a generational category, but as individuals with their own ideas about what life should be like. Stop blaming the kids of the past for the problems of today—and stop feeding the kids of today the lie that their future has already been scripted.
>
> (Willets and Berstow, 2019, n.p.)

Climate victims

Human vulnerability results from the erosion of individual resilience over time, that is, the ability to absorb shocks, self-organise and to adapt (Adger, 2006). The extent to which climate change will affect individual older adults will depend on their age, that is, whether they are in a younger category of older people (age 60–74 years) or an older category (age 75+ years). It will also depend on their sex, health and functional status; wealth and income; seasonal exposure level; and the magnitude of the impact. An individual's resilience to adapt to climate change is determined by both the availability of assets (e.g. amount and quality of knowledge, physical and financial capital, social relations and networks) and access to services (e.g. transport, communication, social support, emergency relief and recovery). However, as a group, older people tend to be more affected by extreme weather events, such as heat waves, tropical cyclones and flooding.

Older people are vulnerable to extreme heat (Harper, 2019). Those who live alone, have pre-existing illnesses, such as cardiovascular disease, are immobile, or suffer from a mental illness, may be particularly disadvantaged (McGregor et al., 2007). Cheng et al. (2018) undertook a study of temperature and deaths among the Australian older population (aged ≥75 years) in five large cities (Adelaide, Brisbane, Melbourne, Perth and Sydney) over the period 1988-2011. They found significant associations between heatwaves and age-related mortality, with higher death rates in the first few days of a heatwave followed by a lower-than-expected death rate.

There is also growing evidence that older people are more likely to die because of tropical cyclones, also called hurricanes and typhoons, due to mobility difficulties; lack of evacuation help and inappropriate evacuation facilities; and disrupted access to essential health and medical support (Harris, 2014; HelpAge International, 2013). Furthermore, floods have several potential negative effects on the health of older people. In addition to immediate injury and death from floodwater, longer-term impacts on health include the spread of infectious disease and increased likelihood of mental illness. Both are exacerbated by the destruction of infrastructure, homes and livelihoods (Watts et al., 2018). Older people are also vulnerable to the impact of wildfires because of increased risk of health effects for short-term exposure to wildfire smoke and an older person's higher prevalence of pre-existing lung and heart disease. The major indirect adverse effects of climate change include temperature-related illnesses and deaths, water stress, air pollution and vector-borne disease (Watts et al., 2018).

Given the contribution some older people make to GHG emissions, we would expect their concerns about climate to be minimal. On the other hand, their vulnerability to climate impacts might increase their concerns about this issue. In order to obtain some clarity on this, the next section looks at the evidence regarding how concerned older people are about climate change and how much they are willing to engage in climate activism.

Is climate concern and action just for the young?

Climate conscious behaviours may be seen as the domain of the young but a 2019 survey of 4,003 UK adults found that the over 55 age group are not only aware of their carbon impact but are more willing to do something about it when compared to their younger counterparts (aged 16–27 years) (Lewis, 2020). However, an earlier 2018 US Gallup poll found a 'global warming age gap' in belief, attitudes and risk perceptions. US adults under 35 years of age were much more engaged with the problem than those 55 years of age and older (Reinhart, 2018). Younger US citizens were more likely to view climate change as important and/or to express a willingness to engage in climate activism (Ballew et al., 2019). Even so, when it comes to actually contacting government officials to urge them to take action to reduce global heating, millennials (born 1981–1986) were more likely than members of Generation X (born 1965–1980) (8%) to have contacted government officials. Another study by Gray et al. (2019) of age and generational differences in environmental concerns found that younger people were no more concerned about climate change than older people. Nor was there evidence to suggest younger generations have a greater willingness to support climate action when compared to older generations. Furthermore, as people grow older they are more likely to engage in recycling, saving energy and buying green products (Gronhoj and Thogersen, 2009). Hence, while there is mixed evidence, in general, older people seem to be as concerned about the environment as their younger counterparts.

The extent to which this concern and individual lifestyle change translates into collective action is also of interest. An ageing society is producing a substantial number of retired, educated, active people who are living longer, and who have different economic and social resources (Pillemer and Filiberto, 2017). Yet the potential for broad mobilisation of older people in environmental civic engagement and volunteerism to combat climate change is being overlooked. Seniors' knowledge of the local environment, its vulnerabilities and how the community has responded in the past allows them to play a key role in reducing the negative impact of current climate-related disasters. In particular, their knowledge of coping mechanisms can be critical when developing local disaster risk reduction and adaptation plans. Environmental volunteering can promote better integration in later life and confer benefits both to the individual (i.e. physical and psychological well-being) and community (e.g. social cohesion). It can expand social connections and intergenerational contact, cognitive function and physical activity. In a study of individual characteristics in older adults and volunteerism, older volunteers with lower socio- economic status reported greater benefits from volunteering (Morrow-Howell et al., 2010). Volunteering also offers older adults the chance to leave a positive legacy which is a powerful motivator for this age group (Villar, 2012). In 2018/2019, people aged 65–74 years were most likely to volunteer on a regular basis in the UK than any other age group. Over one quarter (28%) volunteered at least once a month, while more than one-third (39%) volunteered at least once a year (NCVO,

2020). Furthermore, the intersection of older people's valuing of legacy and the need for climate action could be a pathway for intergenerational collaboration and progress on climate change.

While the possibilities and potential for older people to be active on climate change is there, barriers to engagement on climate issues also exist. This includes a lack of awareness of information sources and the difficulty in accessing information on actions that are best suited to their personal, social or economic circumstances (Haq et al., 2007). The information available is often perceived as being confusing and contradictory. Removal of these barriers, real or perceived, is key to both better engagement of this age sector and achieving sustained involvement. Ageism can also limit participation in activities for older adults within environmental organisations. Younger environmental volunteers have perceived older people as a homogeneous group regarding their capacities, with older people often being invisible to their younger counterparts (Achenbaum 2008). Other barriers to environmental volunteerism may be the perceived lack of individual effectiveness and/or insufficient time or money to devote to such activities (Haq et al., 2007; AARP, 2008).

Despite these barriers, there is a willingness of older people to take action. As witnessed through surveys (Haq et al., 2007), group discussions (Brown et al., 2010) and recommendations within the older people's manifesto (Green Alliance, 2009), the older generation wants to take part in efforts to reduce their environmental impact.

Grey climate activism

Older climate activism exists throughout the world but is often ignored for better known younger groups, such as the school strikes for the climate (Smee, 2019). For example, the US 'Elders for Climate Action' aims to mobilise elders throughout the US to address climate change. Similarly, 'Gray Is Green' is an online gathering of US older adult Americans aspiring to create a green legacy for the future. In Scandinavia, the Norwegian 'Grandparents Climate Campaign' (GCC) was formed in 2006 by a group of concerned elders. In 2020, this independent grassroots organisation had over 3,600 members (February 2020), with regional groups across Norway. Other similar organisations exist in Canada, including 'For Our Grandchildren' and 'The Suzuki Elders'. In France and Switzerland, there is 'Grand-Parents Pour le Climat' and, in the Netherlands, 'Grootouders voor het Klimaat'. In the UK, 'Grandparents Climate Action' is campaigning for the climate. Extinction Rebellion (XR), the non-violent disobedience group, also includes older protestors (Baynes, 2019). For example, a 91-year-old and an 83-year-old, protesting with XR, have been arrested for blocking roads in the UK (Curzon, 2020).

There are a growing number of older protestors who are using retirement to help the climate movement. The US actress Jane Fonda became a climate activist in her 80s, raising awareness about climate change and the anti-fossil fuel movement (Parker, 2019). British nonagenarian David Attenborough is a broadcaster and naturalist who has also become a climate activist in his later years, stating: '[i]f we don't take action, the collapse of our civilisations and the extinction of much of the natural world is on the horizon' (McGrath, 2018, n.p.). When he joined Instagram in 2020, Attenborough was the fastest Instagram users to acquire one million followers, demonstrating the ability of older people to use social media to engage with younger people (Pygas, 2020).

Older people are often more experienced, with some having protested against nuclear weapons, the Vietnam War, the Iraq War and in relation to a number of other political issues and crises in their lifetimes. A seasoned generation of older people have been arrested in global climate-emergency demonstrations. Tonybee describes them as:

> … [T]he fittest ever cohort of pensioners, not only able to sit down in the street, but to get up again, too. These are the best 'arrestables' – free of children, with pensions. They have no need to worry about damaged CVs and criminal-record checks, and so are model protest material, with the least to lose.
>
> *Toynbee (2019, n.p.)*

The recruitment of older adults to civic engagement and volunteerism around climate change issues offers an opportunity for the environmental movement (Pillemer et al., 2016). A vast resource of retired persons may be open to environmental engagement. However, other than their self-organised groups, as listed above, few organised initiatives for older people have so far been available (Pillemer et al., 2010). This has been due to a lack of political awareness of the contribution older people can make to the environmental movement. If senior environmental volunteering is to reach its full potential, then cultural, income, health and practical barriers need to be removed, and discriminatory perceptions and attitudes need to be challenged (Pillemer et al., 2009).

Conclusion and recommendations

Society is ageing as the climate is changing, posing a significant risk to older people. Growing old in the 21st Century will bring with it the unique challenge of adapting to changing weather patterns caused by a warmer climate. This will require a better understanding of the intersection between population ageing and climate change.

In addition to the social justice issue relating to the vulnerability of some older people, it needs to be acknowledged that it is important to address and overcome the stereotype of this age sector as being incapable of engagement, or as passive or disinterested. They remain a 'missing voice' in the environmental debate (Green Alliance, 2009). With the rise in youth climate activism and the anger directed at older generations for the state of the planet, it is easy to assume that older people do not care about climate change. The contribution of older people to climate action can be harnessed by providing opportunities for older people to gain knowledge and become climate movement leaders. Improving and maintaining older people's engagement and action on climate change could be valuable at several levels. In addition to increasing the probability of changing individual behaviours, it responds to the perceived lack of involvement and engagement of older people in collective action. It can also provide a basis for better understanding attitudes to climate change and associated environmental issues.

The views of older people need to be incorporated into policy discussions from a grassroots level to an international level to highlight the impact of climate change on health as people grow older. Enabling older people to be part of the conversation can work to shape decision-making to reflect their needs and address climate change issues. Climate change policy and activism must, therefore, utilise both older and younger people (Curzon, 2020). The older population is not homogeneous, and therefore 'one-size-fits-all' approaches to engagement are unlikely to work (Smyer, 2017). The environmental movement can enable older people to take part in climate activism by ensuring that the organisation's location is accessible and providing a range of volunteer jobs for people of different interests and physical abilities. This includes providing transport when needed and offering daytime activities (The Fifth State, 2017).

Environmental organisations can also take the following steps to engage older people on climate change issues (Haq et al., 2010):

- *Abandon old stereotypes*: Forget negative stereotypes that view older people as being incapable, passive or disinterested. People are living longer due to better health care and diets. This means they are remaining active in later life. Chronological age is no longer an indicator of how an individual will behave in relation to climate change or cope with climate impacts.
- *Get to know your target audience*: Socio-demographic data sets can obscure differences between distinct groups of older people. Get to know your target audience and develop and refine your understanding of differences that exist within the age group. Rather than just using attitude surveys, take into account people's underlying values and their actual behaviour.
- *Use trusted brands*: Information sources, messengers and 'brands' that older people trust, such as charities and established local community groups, should be used when engaging with older people on climate change and green issues.
- *Use peer-to-peer communication*: Older people are more likely to engage with ideas if they are presented by people they know and trust. Existing social networks can be valuable in disseminating ideas, allowing people to share tips and even compare their progress with that of others.
- *Use positive messages*: Positive messages to inspire action should be used, rather than messages of guilt and fear that risk promoting inaction. These should suggest specific, accessible forms of action that older people can adopt.
- *Use the right 'frames'*: A good understanding of the values and priorities of the target audience is needed. Information should be made interesting and meaningful. For older people, this could mean using language and images that draw on values such as thrift, intergenerational justice, legacy thinking and doing their bit for the community.

- *Show real-life examples:* Good examples are important, as they reassure people that they are not alone in taking action, and provide ideas and inspiration. One way to do this is through the institutions and buildings older people come into contact with, such as community centres, churches, shops, hospitals and residential homes. Another is the possibility that adopting environmentally friendly practices might also be financially beneficial.
- *Develop an inclusive dialogue:* Engage people in a dialogue about what works and show that their experience is recognised and valued. Sometimes consultation processes rely on representatives of service providers ('gatekeepers') to speak on behalf of older people. While they may have valuable insights, they do not represent older people's perspectives, and their voices should not be the only ones to be heard.
- *Maximise participation:* To maximise input and involvement, promote discussion and debate rather than lecturing and, where possible, provide adequate financial reimbursements for transport or expenses to support people with a disability that may limit their participation. Timing, formats and safety of consultations are also important considerations.
- *Ensure the setting is right for change:* The engagement of older people must be seen as part of a whole-system change, which includes regulatory, financial and infrastructural policies that promote the transition to a low carbon economy. As part of this, facilities must be provided to enable greener lifestyles, taking into account the different needs and capabilities of older people.

References

AARP (2008) 'More to give: Tapping the talents of the baby boomer, silent and greatest generations', American Association of Retired Persons, Washington, DC.

Achenbaum, W.A. (2008) 'From "Green Old Age" to "Green Seniors": A synoptic history of elders and environmentalism', *Public Policy and Aging Report*, vol 18, no 2, pp. 8–13.

Adger, W.N. (2006) 'Vulnerability', *Global Environmental Change*, vol 16, no 3, pp. 268–281.

Allan, L.J. and Johnson, J.A. (2009) 'Undergraduate attitudes toward the elderly: The role of knowledge, contact and aging anxiety', *Educational Gerontology,* vol 35, pp. 1–14.

Baldwin, C., Matthews, T., Byrne, J. (2020) 'Planning for older people in a rapidly warming and ageing world: the role of urban greening.' Urban Policy and Research, Vol 38, no 3, pp 199–212.

Ballew, M., Marlon, J., Rosenthal, S., Gustafson, A., Kotcher, J., Maibach, E. and Leiserowitz, A. (2019), 'Do younger generations care more about global warming?', Yale Program on Climate Change Communication. Yale University and George Mason University, New Haven, CT.

Baynes, C. (2019) 'Extinction Rebellion: 91-year-old among activists arrested after elderly protesters "glue" themselves to Dover roads', *The Independent*, 21 September.

Biggs, S., Phillipson, C., Leach, R. and Money, A. (2007) 'The mature imagination and consumption strategies age and generation in the Development of a United Kingdom Baby Boomer Identity', *International Journal of Ageing and Later Life*, vol 2, no 2, pp. 31–59.

Bouzid, M., Hooper, L. and Hunter, R. (2013) 'The effectiveness of public health interventions to reduce the health impact of climate change: A systematic review of systematic reviews', *PLoS One*, vol 8, no 4, e62041. https://doi.org/10.1371/journal.pone.0062041

Bravo-Segal, S. and Villar, F. (2020) 'Older people representation on the media during COVID-19 pandemic: a reinforcement of ageism?', *Revista Española de Geriatría y Gerontología*, vol 55, no. 5, pp. 266–271.

Brennan, M., O'Keeffe, S. T., Mulkerrin, E. C. (2019) 'Dehydration and renal failure in older persons during heatwaves-predictable, hard to identify but preventable?' Age and Ageing, Vol 48, no 5, pp 615–618.

Brown, D., Hamid, S., Morrison, A., Williams P. (2010) 'Retired and Senior Volunteer Programme Climate Change and the Over 50s', Swindon Borough Council Community.

Buchs, M., Bahaj, A., Blunden, L., Bourikas, L., Falkingham, J., James, P., Kamanda, M. and Wu, Y. (2018) 'Sick and stuck at home – how poor health increases electricity consumption and reduces opportunities for environmentally friendly travel in the United Kingdom', *Energy Research and Social Science,* vol 44, pp. 250–259.

Chancel, L. (2014) 'Are younger generations higher carbon emitters than their elders? Inequalities, generations and CO_2 emissions in France and in the USA', *Ecological Economics*, vol 100, pp. 195–207.

Cheng, J., Xu, Z., Bambrick, H., Su, H., Tong, S. and Hu, W. (2018) 'Heatwave and elderly mortality: An evaluation of the death burden and health costs considering short-term mortality displacement', *Environment International*, vol 115, pp. 334–342.

Curzon, H. (2020) 'Not us and them – The voice of older people in the climate crisis', International Longevity Centre, https://ilcuk.org.uk/not-us-and-them/.

Donizzetti, A.R. (2019) 'Ageism in an aging society: The role of knowledge, anxiety about aging and sterotypes in young people and adults', *Environmental Research and Public Health*, vol 16, no 1329.

Doyle, P., Kelly, I., O'Neill, D. (2019) 'Older people: Canaries in the coal-mine for health effects of climate change.' Irish Medical Journal, vol 112, no 10, pp 1015.

Estiri, H. and Zagheni, E. (2019) 'Age matters: Ageing and household energy demand in the United States', *Energy Research and Social Science*, vol 55, pp. 62–17.

Frumkin, H., Fried, L. and Moody, R. (2012) 'Aging, climate change, and legacy thinking', *American Journal of Public Health*, vol 102, no 8, pp. 1434–1438.

Gray, S.G., Raimi, K.T., Wilson, R. and Arvai, J. (2019) 'Will millennials save the world? The effect of age and generational differences on environmental concern', *Journal of Environmental Management*, vol 242, pp. 394–402.

Green Alliance (2009) 'Greener and Wiser. An older peoples manifesto on the environment', Green Alliance, London.

Gronhoj, A. and Thogersen, J. (2009) 'Like father, like son? Intergenerational transmission of values, attitudes, and behaviours in the environmental domain', *Journal of Environmental Psychology*, vol 29, no 4, pp. 414–421.

Hamza, N. and Gilroy, R. (2011) 'The challenge to UK energy policy: An ageing population perspective on energy saving measures and consumption', *Energy Policy*, vol 39, pp. 782–789.

Haq, G., Minx, J., Whitelegg, J. and Owen, A. (2007) 'Greening the greys: Climate change and the over 50s', Stockholm Environment Institute, Sweden.

Haq, G., Whitelegg, J. and Kohler, M. (2008) 'Growing old in a changing climate: Meeting the challenges of an ageing population and climate change', Stockholm Environment Institute, Sweden.

Haq, G., Brown, D. and Hards, S. (2010) 'Older people and climate change: The case for better engagement', Stockholm Environment Institute, Sweden.

Harper, S. (2019) 'The convergence of population ageing with climate change', *Journal of Population Ageing*, vol 12, pp. 401–403.

Harris, C. (2014) 'Disaster resilience in an ageing world', HelpAge International, London.

HelpAge International (2013) 'Older people disproportionately affected by Typhoon Haiyan', HelpAge International, London. www.helpage.org/newsroom/latest-news/older-people-disproportionately-affected- by-typhoon-haiyan.

Hitchings, R., Collins, R. and Day, R. (2015) 'Inadvertent environmentalism and the action-value opportunity: Reflection from studies at both ends of the generational spectrum', *Local Environment*, vol 20, no 3, pp. 369–385.

Hitchings, R., Venn, S. and Day, R. (2018) 'Assumptions about later-life travel and their implications: pushing people around?' *Ageing and Society*, vol 28, no 1, pp. 1–18.

IPCC (2014) 'Climate Change 2014: Synthesis report. Contribution of working groups I, II and III to the fifth assessment report', Intergovernmental Panel on Climate Change, Geneva, Switzerland.

IPCC (2018) 'Summary for policymakers', in *Global Warming of 1.5°C*. Intergovernmental Panel on Climate Change, Geneva, Switzerland.

Karpf, A. (2020) 'Don't let prejudice against older people contaminate the climate movement', *The Guardian*, 18 January.

Lewis, S. (2020) 'Generation woke? Over 55s most likely to recycle, study shows', *Aviva*, 13 February.

Malak, M. A., Sajib, A. M., Quader, M. A. and Anjum, H. (2020) 'We are feeling older that our age: vulnerability and adaptive strategies of ageing people to cyclones in coastal Bangladesh.' International Journal of Disaster Risk Reduction, Vol 48, 101595

McGrath, M. (2018) 'Sir David Attenborough: Climate change "our greatest threat"', *BBC News*, 3 December.

McGregor, G.R., Pelling, M., Wolf, T. and Gosling, S. (2007) 'The social impact of heat waves', Environment Agency, Science Report – SC20061/SR6, Bristol.

Moody, R. (2017) 'Elders and climate change: No excuses', *Public Policy and Aging Report*, vol 27, no 1, pp. 22–26.

Morrow-Howell, N. (2010) 'Volunteering in later life: Research frontiers', *Journals of Gerontology, Psychological Sciences and Social Sciences*, vol 65, no 4, pp. 461–469.

NCVO (2020) 'What are the demographics of volunteers?' https://data.ncvo.org.uk/volunteering/demographics/

NPR (2019) 'Transcript: Greta Thunberg's Speech at the U.N. Climate Action Summit' www.npr.org/2019/09/23/763452863/transcript-greta-thunbergs-speech-at-the-u-n-climate-action-summit?

Okada, A. (2012) 'Is an increased elderly population related to decreased CO_2 emissions from road transportation?', *Energy Policy*, vol 45, pp. 286–292.

O'Neil, B.C., Liddle, B., Jiang, L., Smith, K.R., Pachauri, S., Dalton, M. and Fuchs, R. (2012). 'Demographic change and carbon dioxide emissions', *The Lancet*, vol 380, pp. 157–164.

Otto, S. and Kaiser, F.G. (2014) 'Ecological behavior across the lifespan: Why environmentalism increases as people grow older', *Journal of Environmental Psychology,* vol 4, pp. 331–338.

Parker, R. (2019) 'Jane Fonda arrested a fifth time while protesting in Washington, D.C.', *The Hollywood Reporter*, 20 December.

Pillemer, K., Fuller-Rowell, T.E., Reid, M.C. and Wells, N.M. (2010) 'Environmental volunteering and health outcomes over a 20-year period', *The Gerontologist*, vol 50, no 5, pp. 594–602.

Pillemer, K., Wells, N.M., Meador Rhoda, H., Schultz, L., Henderson Jr., C.R. and Cope, M.T (2016) 'Engaging older adults in environmental volunteerism: The Retirees in Service to the Environment (RISE) program', *The Gerontologist*, vol 57, no 2, pp. 367–375.

Pillemer, K. and Filiberto, D. (2017) 'Mobilizing older people to address climate change', *Public Policy and Aging Report*, vol 27, no 1, pp. 22–26.

Pillemer, K., Wagenet, L., Goldman, D., Bushway, L. and Meador, R. (2009) 'Environmental volunteering in later life: Benefits and barriers', *Generations*, no 4, pp. 58–63.

Pygas, M. (2020) 'David Attenborough breaks record as fastest Instagram user to 1 million followers', Greenmatters. www.greenmatters.com/p/david-attenborough-instagram-one-million-followers.

Reinhart, R.J. (2018) 'Global warming age gap: Younger Americans most worried', Gallup. https://news.gallup.com/poll/234314/global-warming-age-gap-younger-americans-worried.aspx.

Romanach, L., Hall, N. and Meikle, S. (2017) 'Energy consumption in an ageing population: exploring energy use and behaviour of low-income older Australians', *Energy Procedia*, vol 121, pp. 246–253.

Smee, B. (2019) 'Warming world gets older, wiser, richer activists hot under the collar', *The Guardian*, 26 January.

Smyer, M.A. (2017) 'Greening gray: Climate action for an aging world', *Public Policy and Aging Report*, vol 27, no 1, pp. 4–7.

Sánchez-Gonzalez, D. and Chávez-Alvarado, R. (2016) 'Adjustments to physical-social environment of the elderly to climate change: Proposals for environmental gerontology', in D. Sánchez-Gonzalez and R. Chávez-Alvarado (eds.), *Environmental Gerontology in Europe and Latin America*. Springer, Cham, Switzerland.

Sands, K. (2018) *The Ageless Way: Illuminating the Story of Our Age*. 2nd edn. Broad Minded, Roxbury.

Tasdemir, N. (2020) 'Young group identification and motives as predictors of ageism, aging anxiety, and life satisfaction', *The Journal of Genetic Psychology*, vol 181, no 5, pp. 375–390.

The Fifth State (2017) 'Grey and green: How older people are the perfect climate allies', *The Fifth State*, 30 March.

Thunberg, G. (2020) 'At Davos we will tell world leaders to abandon the fossil fuel economy', *The Guardian*, 10 January.

Toynbee, P. (2019) 'How older people became the heroes of Extinction Rebellion', *The Guardian*, 15 October.

Trendell, A. (2019) 'Billie Eilish: "Greta Thunberg is paving the way. Hopefully the old people start listening to us so we don't all die"', *NME*, 16 December.

UN (2019) 'Ageing', www.un.org/en/sections/issues-depth/ageing/.

UNDESA (2019) 'World population prospects – Highlights', United Nations Department for Economic and Social Affairs, New York.

Unruh, C. (2019) 'We want to fight climate change with you, not against you', Intergenerational Foundation. www.if.org.uk/2019/10/28/we-want-to-fight-climate-change-with-you-not-against-you/.

Venn, S., Burningham, K., Christie, I. and Jackson, T. (2017) 'Consumption junkies or sustainable consumers: considering the shopping practices of those transitioning to retirement', *Ageing and Society*, vol 37, no 1, pp. 14–38.

Villar, F. (2012) 'Successful ageing and development: The contribution of generativity in older age', *Ageing Society*, vol 32, no 7, pp. 1087–1105.

Willetts, D. and Berstow, J. (2019) 'Have the old stolen young people's futures? Our contributors debate', *Prospect Magazine*, 7 October.

Wei, T., Zhu, Q. and Glomsrod, S. (2018) 'How will demographic characteristics of the labor force matter for the global economy and carbon dioxide emissions?', *Ecological Economics*, vol 147, pp. 197–207.

Watts, N., Amann, M., Aveb-Karlsson, S., Belesova, K., Bouley, T., Bovkoff, M. et al. (2018) 'The Lancet countdown on health and climate change: From 25 years of inaction to a global transformation for public health', *Lancet*, vol 391, no 10120, pp. 581–630.

Wu, J., Snell, G. and Samji, H. (2020) 'Climate anxiety in young people: A call to action', *Lancet Planet Health*. https://doi.org/10.1016/S2542-5196(20)30223-0

Zagheni, E. (2011) 'The leverage of demographic dynamics on carbon dioxide emissions: Does age structure matter?' *Demography*, vol 48, pp. 371–399.

Zaval, L., Markowitz, E.M. and Weber, E.U. (2015) 'How will I be remembered? Conserving the environment for the sake of one's legacy', *Psychological Science*, vol 26, no 2, pp. 231–236.

9

YOUTH, CLIMATE AND ENVIRONMENTALISM

Benjamin Bowman, Karen Bell and Becky Alexis-Martin

Introduction

In this chapter, we focus on the activism of children. Because they are 'highly marginalized both economically and socially', and because their lives are 'structured by domination', children are accurately to be considered subaltern activists (Pulido, 1997, p. 25). Their activism, then, is more than a struggle to gain purchase on policymaking. The activism of children is also, by its nature, a claim on the right to agency in conditions of social, economic and political inequality that deny young people, and especially children, the right to a voice and the power to take action.

Youth is a category of marginalisation that intersects with other inequalities, including, but not limited to, racialised injustice, economic inequality, gendered inequality and the unequal burden of environmental degradation and harm in different locations across the world. The prominence of young people in environmental politics includes many well-known icons: prominent activists like the Swedish teenager Greta Thunberg, the Marshallese poet-activist Kathy Jetñil-Kijiner and the Canadian activist Severn Cullis-Suzuki, all of whom have addressed the United Nations on behalf of young people. Young people have also been prominent in environmental movements and 'green' causes since the emergence of the modern Western environmental movement in the 1960s and 1970s. Yet, young people are marginalised as political agents. The marginalisation of children, in particular, almost always includes the inability to vote, unequal access to democratic fora, transitional status as 'citizens-to-be' or 'apprentice adults' as opposed to full recognition, and so forth.

Young environmentalist activism is not monolithic, but it is characteristically oriented towards the environmental justice approach (Bowman, 2020). 'Environmental justice' is a term with roots in the work of Black and Latino activists in the United States (US) in coordination with the civil rights campaigns of the 1950s and 1960s, as well as indigenous resurgence and sovereignty movements (Pulido, 1997, p. 25). Young activism as a form of environmental justice activism can challenge mainstream concepts of environmentalist activism because young people frequently address environmental issues at their intersection with the enduring damage of colonial regimes, economic exploitation, the fight for land rights and outdoor spaces, everyday health and safety, and structural racism (Bowman, 2020).

Because young people experience marginalisation, domination and oppression, the environmentalism of young people around the world has historically been typified by what Orlando Fals Borda (cited in Haynes, 1999, p. 223) called an attraction to projects 'to bring about a new ethos, a better kind of society and social relations in which unity may coexist with diversity'. This is in a context of 'the disproportionate racialized and classed impacts of environmental damage' (Pulido, 1997, p. 25). The marginalisation of young people often includes their marginalisation from mainstream environmentalism, especially at times when the mainstream movement 'artificially compartmentalizes people's troubles' in order to render environmentalism by marginalised groups 'not environmental' (Pulido, 1997, p. 25). For instance, the condensing of young activism, with its rich history and complex politics of environmental justice, to familiar, 'narrowly constructed, technocratic and dehistoricized' (Hamilton Faris, 2019, p. 77) goals and slogans like 'listen to the science'.

Children's activism is complex. It tends to be framed in complex ways, and generally speaking, it is framed in ways that uphold and recognise the connections between environmental justice and global economic, social and political justice. Children demonstrate an 'expanded toolbox for political action' (Bowman, 2019) that includes 'do it ourselves' approaches to everyday politics (Pickard, 2019), 'engaged scepticism' (O'Toole et al., 2003) and 'lifestyle politics' (Vromen et al., 2015). At the same time, the complexity of children's environmental activism is 'typically lost in adult reporting' which continues to privilege top-down, adult-centred ontologies and 'engagement framing' that recasts young climate activism as mere displays of engagement (Bowman, 2020, p. 2). In this chapter, we discuss the importance, and marginalisation, of children as environmental activists. We focus on children, but reflect at the end of the chapter on young people's activism more widely.

The importance of considering young people

There are a number of reasons that it is important to consider the inclusion and diversity of young people in relation to climate and environmentalism. Because so much of the Earth has already been harmed, young people will spend most of their lives in environmentally degraded situations. Their security, incomes and mental health will be hugely impacted by climate change alone (Eskenazi et al., 2020). A number of studies have already shown how climate change has negatively impacted the physical and mental health of children and young people (e.g. Majeed and Lee, 2017; Sanson et al., 2019). In the Global South, these impacts have been particularly severe, for example, reducing the availability of water and food security. It is estimated that children experience more than 80% of the illness and mortality attributed to climate change (Currie and Deschênes, 2016). Children are more impacted in terms of health-related environmental issues, such as 'increased heat stress; decreased air quality; altered disease patterns of some climate-sensitive infections; and food, water, and nutrient insecurity in vulnerable regions' (Council on Environmental Health, 2015, p. 992). Young people are facing the possibility of a poorer, less healthy and more conflict-prone world. They are also marginalised in other ways, such as through poverty, gender, ethnicity, queerness or as disabled people. It is also important to consider young people, as this chapter discusses, because they have been environmental leaders and, despite their limited democratic power, have made a major impact on environmentalism and improved its inclusiveness. Children and young people have been engaged in climate activism from the national level to influencing international policy (Olson, 2016, p. 81; YOUNGO, 2019).

Research on young people's environmental concerns

The major limitation in academic research regarding young people's environmentalism is that much of it remains dominated by adult-centred conceptual frameworks that limit, or even fully obscure, what environmentalism means for young people. These limitations can yield research that has been criticised as 'over-structured investigations of the attitudes of school children to adult political concepts' (Crick, 1999, p. 342), and poorly serve 'young people's own uncertainty and ambiguity in describing what "counts" as political action beyond the formal and adult-centric definitions which they are likely to have been exposed to in their life course' (Wood 2014, p. 214). For example, while social movement studies typically define motivations to protest according to a binary epistemology that divides instrumental goals from expressive goals, children's political practice is more complex, and they 'take up more individualized and everyday practices in efforts to shape society' (Harris et al., 2010, p. 28).

The dominant epistemologies in academic research are adult centred. Adult-centred epistemologies tend to divide and typologise young people's attitudes, activities, subjectivities and so forth into coherent, adult-oriented categories, and especially to divide the political from the personal and the public sphere from the private. Such epistemologies are a poor fit for young people's politics, which is more often hybrid and shifting. Young people's political subjectivities, like young people themselves, are liminal, and they straddle the threshold of the public and the private.

Youth narratives on environmentalism have taken diverse forms around the world (Thomas et al., 2019). Youth environmental politics takes different forms according to location and, perhaps also, age. Lee et al. (2020), synthesising research from the fields of educational science, psychology, geography and the broader environmental social sciences, looked at what studies (1993–2018) revealed about children and adolescents' beliefs and concerns about climate change and their views on its causes, impacts and solutions. They found that, among young people, levels of belief, concern and willingness to take action were lower in the US, United Kingdom (UK) and Australia than in the other countries they looked at. They also found that younger children were more willing to take action on climate change compared to older children (e.g. Malandrakis et al , 2011; Ambusaidi et al., 2012). They were also more willing to use environmentally friendly transport compared to older children (Chhokar et al., 2011; Skamp et al., 2009; Ambusaidi et al., 2012). The authors (Lee et al., 2020) felt that this could be explained by the fact that younger children usually have fewer opportunities to make decisions about transport and other environmental issues so they can be ideological, without having to actually carry through the action. Older children will be more aware of the lack of convenience associated with more environmentally friendly choices. It is also suggested that reduced concern and willingness to act may reflect coping strategies, where young people feel powerless to make a difference (Ojala, 2012a, 2012b).

Young disillusionment with institutional politics, and a sense among many young people that institutional modes of political action (and especially voting) are ineffective, goes hand-in-hand with the hybrid and liminal nature of young people's political subjectivities. Young people tend to take action on a case-by-case basis in issues that are meaningful in their everyday lives. They are 'project-oriented, and want to deal with common concerns concretely and personally rather than abstractly and ideologically' (Bang, 2005, p. 162). For this reason, environmental concerns among young people tend to be framed 'within the context of other existential (economic and social) challenges' including economic precarity (Sloam 2020, p. 2). It is relevant − and important − to point out that young people tend to perceive environmental concerns to be closely linked to economic hardship, pollution, access to green spaces, self-determination, fairness and justice. These issues have everyday meaning and can support concrete, personal action. Yet, in a similar way to the environmental concerns of other marginalised groups, young people often find that mainstream environmentalism, and mainstream adult scholarship, 'artificially compartmentalizes people's troubles' in order to render minority environmental activism 'not environmental' (Austin and Schill 1991, p. 72).

Young people as activists

Many studies have asserted low levels of political participation and environmental activism among young people, but it is generally the case that such studies enforce adult terms and adult concepts of politics, environmentalism and activism on young people. A broad academic literature exists that presupposes adult-centric approaches and enforces, for example, methodological divisions between political motivations and self-expressive motivations (Bowman, 2019). Such literature frequently claims that young people are disengaged from civic action (Costanza-Chock, 2012) and − it must be said, rather confusingly − it is often claimed that young people 'fall short of offering concrete solutions' (as in Han and Ahn, 2020) or, alternatively, that young people are too limited by their focus on concrete solutions (Evensen, 2019). The numerous critiques of young activism, and particularly the climate strike movement, tend to fail to recognise the marginalisation of young people including children from the civic realm, from democratic processes, and their marginalisation in society, generally. Criticisms of young activism also tend not to involve young people themselves, nor do they frequently cite youth-authored texts like the Lausanne Declaration of the Fridays for Future movement, which outlines concrete solutions as well as commitments to social justice, global equality and so forth (Fridays for Future, 2019).

Activists, at all ages, struggle to be recognised as 'agents of democracy, not pathological deviants' (Haste et al., 2017, p. 4). Child activists, because they are children, must push against their civic exclusion as children whose role is to engage with politics, not to change it; to have a voice, but not power; to sustain democratic systems by becoming socialised to them, not to challenge the status quo. Political action by children challenges the concept of the civic, since young people's democratic participation is directed towards 'consent, cohesion and loyalty, rather than contestation and dissent' (O'Brien et al., 2018, p. 5). By no coincidence, young people's environmentalist action is often shaped by everyday experience and their participation may be issue-based, local and on a case-by-case basis (Sloam, 2020).

Epistemologically speaking, the search by adults to identify children's participation as somehow 'activist' tends to privilege modes of participation that cohere with adult institutions, are loyal to adult-led groups and are easily comprehensible to adult policymakers, adult activists and the mainstream adult environmentalist movement. The renewal of young politics is celebrated in these terms. For example, in the UK, in a reversal of a historic decline in youth electoral participation, many young people joined and became influential in the Labour Party when under the leadership of Jeremy Corbyn (Pickard, 2018; Young, 2018). Other prominent forms of young activism have included taking out lawsuits against fossil fuel companies and governments (Parker, 2019). For example, in 2015, young people from Oregon, US, filed a lawsuit against the federal government and the fossil fuel industry. They argued that their government's failure to address climate change and protect resources had impacted their life, liberty and property (Our Children's Trust, n.d.). Lawsuits along similar lines have also been taken out in Colombia, Pakistan and other countries of the Global South (Parker, 2019). The UN Development Program has financed a number of youth-led environmental projects globally (UN Development Program, 2015).

One of the most important manifestations of their concern and willingness to take action has been the youth mobilisation against climate change, but there have been numerous prior youth movements throughout history. For example, young people recently mobilised against gun violence in the US. They have also played an important role in the feminism, anti-war, labour and anti-racism movements (Costanza-Chock, 2012; Ouellett, 1996).

What stands out about the climate strike movement may be the way that school strikes take place across the conceptual and epistemological boundary of child activism because they disrupt adult institutions, align with adult-led groups and present an easy-to-recognise form of civil disobedience, which is strike action and the withdrawal of labour. At the same time, the climate strikes offer children the opportunity to dissent and disrupt, to express themselves, to 'do it ourselves' (Pickard, 2019) and to organise their own movement led by young people for young people.

Youth climate strike

School strikes are familiar as a tool for children to take action through civil disobedience, from the strike action of young people during the Soweto uprisings, to school strikes in the UK against the invasion of Iraq in 2003, to the strikes of young people for stricter controls on gun ownership as part of the March for our Lives movement in the US, among many examples. Children's strikes from school for action on climate change were timed, originally, to coincide with the 2015 UN Conference of the Parties (COP21) in Paris. The current global movement is often credited to Greta Thunberg, who has been a focus of media attention since she began her protests, 'Skolstrejk för klimatet' ('School strike for climate') in 2018. Since she was 15 years old, the Swedish climate activist has spearheaded the campaign, variously known as 'FridaysForFuture', 'Youthforclimate' and 'Youthstrike4climate'. The strike became international when, on 15 March 2019, young people from over 100 countries walked out of their classes and joined protests demanding that their governments take action to prevent further climate change. There have been regular protests since then, though the coronavirus (COVID-19) situation has prevented their continuance in physical manifestations, and movement activities are mostly taking place online. It is estimated that 6 million young people globally took action during the week commencing 20 September 2019 (Taylor et al., 2019).

The school students argued that it will be their generation that will be most affected by a lack of climate change mitigation (Warren, 2019). They have emphasised that they will be faced with the escalating problems created by climate change and that their future lives will be negatively affected by its impact. Greta Thunberg argued for recognition of 'the science' and the movement received the support of many scientists and academics (Hagedorn et al., 2019; Scientists for Future, 2019).

Greta Thunberg stated at the Davos Forum:

> And why is it so important to stay below 1.5° Celsius? Because even at 1° people are dying from climate change because that is what the united science calls for, to avoid destabilizing the climate so that we have the best possible chance to avoid setting off irreversible chain reactions.
>
> *(Thunberg, 2019a)*

The young people have highlighted that the international community has failed to act on its commitments to reduce emissions.

It is suggested that participation in the school strikes facilitates increased 'connectedness to nature' and learning about climate change, factors shown to increase ecological behaviour (Otto and Pensini, 2017). It has been argued that '…a motivation to behave in an ecologically friendly manner is formed in childhood and has the potential to be lifelong, so it is possible that participating in the strikes could have a long-lasting effect on the ecological behavior of this cohort' (Lee et al., 2020, p. 14).

Although other young people had made similar public comments to that of Thunberg in the past, the power of social media allowed her to have a much greater impact. For example, in 1992, 12-year-old Severn Cullis-Suzuki, the founder of the Environmental Children's Organization, spoke at the Rio Earth Summit (Cullis-Suzuki, 1992). Thunberg began her 'School Strike for Climate' in late August 2015 outside the Swedish parliament (BBC, 2020). She used a number of social media outlets such as Instagram, Twitter, Facebook and YouTube – and her global peers began sharing hashtags, such as #FridaysForFuture (FFF) and #Climatestrike. In December 2018, she gave a speech at the UN climate change conference, COP24.

Responding to the youth surge, the UN hosted its first Youth Climate Summit on 21 September 2019. The young climate attendees discussed ways to meet the commitments in the Paris Agreement and demanded stronger action to mitigate climate change (UN, 2020). Thunberg travelled to the summit on a solar-powered sailing boat. On 20 September, before the UN youth climate summit, young people in at least 117 countries protested against the inadequate action taken by world leaders, with an estimate 4 million people taking to the streets that day (Alter et al., 2019).

At the UN Youth Climate Summit, Thunberg and 14 children from Argentina, the Marshall Islands, France, Germany and the US made a formal complaint that the failures of countries to address the climate crisis violated the UN Convention on the Rights of the Child (Milman, 2019). A Fijian climate activist, Komal Kumar, said that young people should be engaged in the design of adaptation plans. She described her generation as 'living in constant fear and climate anxiety' (UN News, 2019, n.p.).

Young people are very aware of their limited voice in many so-called democracies. For example, at a speech to the European Union Parliament in April 2019, Greta Thunberg said:

> The EU elections are coming up soon, and many of us who will be affected the most by this crisis, people like me, are not allowed to vote. Nor are we in a position to shape the decisions of business, politics, engineering, media, education, or science. Because the time it takes for us to educate ourselves to do that simply does no longer exists, [sic] and that is why millions of children are taking it to the streets, school striking for the climate to create attention for the climate crisis. You need to listen to us, we who cannot vote.
>
> *(Thunberg, 2019b)*

The climate strike movement, we argue, is likely to prove durable. Han and Ahn (2020) point out that while youth environment movements are not entirely new, the global scale of the youth mobilisation on environmental issues was unprecedented.

Bowman (2020) also writes that the scale of the climate strike movement must be understood in the context of a long history of young environmentalist activism. The climate strikes are highly connected through social media and other digital tools (Boulianne et al., 2020) and are likely to develop further in future.

Impact of young people as environmental activists

Although it is difficult to assess the impact of social movements, there are a number of indications that the recent youth action has had a huge impact on understanding and action in relation to climate change. It is clear that it has increased a sense of urgency, highlighted issues of justice and cultivated youth leadership. Primarily, it has attracted an enormous amount of media interest, and thereby public attention. Meltwater, a media monitoring company (Munawar, 2019), found more than 7.5 million social media mentions of Thunberg in the week before the UN Climate Action Summit, from 20 September 2019 to 26 September 2019, and news outlet mentions of her on 93,800 occasions. Most of these mentions were aimed at increasing climate change awareness and urging people to join social movements to put pressure on policymakers to take mitigating action. *Collins Dictionary* declared 'climate strike' as the word of the year (2019) because of a 100-fold increase in the use of the word. It was defined as a 'form of protest in which people absent themselves from education or work in order to join demonstrations demanding action to counter climate change' (Guardian, 2019, n.p.). Han and Ahn (2020) stated that these climate-related expressions gained global recognition in that year mainly as a result of the global youth mobilisation.

Han and Ahn (2020) also noted that the youth climate movements involved other social movements, including labour organisations, teachers and other environmental groups. For example, in the US, the American Federation of Teachers supported the students in relation to student absences, logistics and endorsements (Myers, 2019). Other groups supporting the US school strikes included the Sunrise Movement, Zero Hour, OneMillionOfUs, 350.org and the National Children's Campaign (Myers, 2019). A number of trade unions supported the strikes around the world (Han and Ahn, 2020). For example, in Germany, the large service sector union, Verdi, urged its members to join the strike in September 2019.

Han and Ahn (2020) argued that the 2018–2019 youth-led climate movements can be said to have had an impact on the climate change policies of some countries. For example, following the UN Climate Action Summit, Angela Merkel, Chancellor of Germany, stated that coal mining would end in Germany by 2038, and Emmanuel Macron, president of France, declared that France would not make trade deals with countries that had not endorsed the Paris Agreement (Milman, 2019). Ten governments, including Sweden, Chile and Spain, signed the Declaration on Children, Youth, and Climate Action (Declaration on Children, Youth, and Climate Action, 2019) at the COP25 in Madrid. This declaration recognised the critical role of children and youth as agents of change. It included a commitment to taking action on the promotion of youth rights, including the right to a healthy environment; and the promotion of youth participation in climate governance. At the UN Youth Climate Summit, Amina Mohammed, the deputy secretary general, stated that the UN had never offered such a visible platform to young people at a political summit, and emphasised how they were drawing worldwide attention to the climate emergency (UN News, 2019).

Solutions

Young climate movements, and especially the climate strike movement as the currently prominent, and most widely recognised, youth-led environmentalist movement, suggest many solutions to environmental issues, such as climate change. What characterises these solutions is that young people tend to be environmental justice activists rather than mainstream environmentalists (Bowman, 2019) and a desire for radical systemic change is frequently 'written between the lines' (Bowman, 2019, p. 296) in young climate activism.

Prominent among the civil disobedience of youth activists is an explicit critique of the unequal distribution of power and resources that underpin the climate crisis. Generally speaking, young climate activists draw attention to the ongoing inefficacy of existing political processes to force governments and fossil fuel companies to take any action. As Greta Thunberg stated in her speech at the 2019 UN Climate Action Summit, the problem is frequently perceived to be one of capitalist politics and economics: '[w]e are in the beginning of a mass extinction and all you can talk about is money and fairy tales of eternal economic growth' (Thunberg, 2019c). Yet, we argue, it is not entirely appropriate to define the movement as anti-capitalist. In its complexity, the climate strike movement resists definition in such terms. The central leadership of the movement has aligned with some concrete goals but, broadly speaking, remains pragmatic about the limited capacity of activists to offer concrete solutions to enormous, systemic problems. The climate strike movement is one that often celebrates the creation of venues to ask questions, as opposed to aligning participants with centrally negotiated and neatly packaged solutions.

The most comprehensive survey of Fridays for Future activists around the world showed that:

> ...[I]n almost every country, student and adult participants are extremely sceptical about relying on companies and the market to solve environmental problems. There are significant differences between countries, and between adults and school students, over stopping climate change through individual lifestyle changes, highlighting that the movement may actually be quite heterogeneous in some regards.
>
> *(Wahlström et al., 2019, p. 5)*

The heterogeneity of the climate strike movement is a feature of the movement, not a flaw. As Pulido (1996) wrote, the activism of those who are marginalised is characterised by the negotiation of multiple positions. In environmentalism, these positions sometimes put young people in situations of what have been called contradictory solidarities (Curnow and Helferty, 2018).

As the organiser, Jen Gobby (2019, p. 245), reflects on the climate strikes, the 'seeds of radical change' are often to be seen in 'their signs and banners [which] read "System Change, Not Climate Change"'. Yet – as Gobby writes – the young people themselves remain bound in intersecting systems of oppression, injustice, damage and harm. A movement which brings young people together across global geographic divides, across local and global inequalities, and despite histories of colonial violence and oppression, environmental racism, and all manner of systemic problems that young people may wish to change, is heterogeneous in nature. Criticisms of the climate strike movement that call out a lack of singular, concrete goals or clear political leadership miss the point that this movement is not a protest movement, calling for piecemeal adjustments to existing political and economic systems in order to render those systems sustainable. It is a movement of young dissent (O'Brien et al., 2018) that would prefer those systems not to be sustained. It is a movement of civil disobedience that disrupts those systems, led by young people, including children, who recognise that 'the climate crisis has made an always uncertain future even more unknowable' (Buckley, 2020) and want to build a better world, not renew their commitment to the current one.

Barriers

While young people face similar barriers to adults in terms of influencing environmental policy, their limited power is compounded by their general lack of political rights in terms of voting and representation. At every level, from the local to the national to the international politics, environmentalists are up against vested interests. If the rich and powerful refuse to 'listen to the science', why would they listen to young people?

Young people lack power in society. While they have moral authority, they are not an adequate match for vested interests. There is also the issue of adultism. While millions of children and young people have taken action on climate change, they must ultimately wait for a response from adults. As Josefsson and Wall (2020) highlight, post-colonialist analyses draw attention to the scenario where marginalised groups are not only oppressed, but also frequently considered 'childlike', irrational, or less than fully 'developed'. They comment:

> The disempowerment of the young, just like the disempowerment of women, is deeply rooted in centuries of biased politics, economics, mores, literature, art, language, and just about every aspect of social life. We suggest that global political culture needs to respond to problems like the climate emergency in part through a kind of broad-based cultural self-critique that actively empowers rather than just passively tolerates historically marginalized voices and experiences.
>
> *(Josefsson and Wall, 2020, p. 14)*

As Spivak (1988, 1999) asked, 'Can the subaltern speak?'. In other words, how can power be challenged by the very same subaltern groups that they silence? It is particularly difficult for children and young people to establish their right to speak on any matter. Young people can be dismissed as troublesome and the youth activists have been brushed aside as immature. For example, after Greta Thunberg, nominated for a Nobel Peace Prize in 2019, tweeted about the murder of indigenous people in Brazil, Jair Bolsonaro called her a 'little brat' (Alter et al., 2019).

Josefsson and Wall (2020) speak of the necessity for the 'empowered inclusion' of children in order for their voices to be heard and for them to have the influence that they deserve. They explain that empowered inclusion is more than just providing children and other marginalised groups with a space in which to have their voices heard, and it also requires global political leaders such as at the UN, global corporations, national governments and NGOs, to make self-critical responses and transform their own understandings.

Young people need to be recognised as possessing valid skills and knowledge and to be meaningfully engaged in deliberative processes (Checkoway, 2011; Thomas et al., 2019; Amponsem et al., 2019). They have had their own parallel forums and formal constituency status at international environmental and climate change fora, such as YOUNGO (Children and Youth Constituency to the United Nations Framework Convention on Climate Change, 2019) and the SDGs (United Nations Major Group for Children and Youth, 2017). But participation is no guarantee of having influence, and it is noted that these fora can be tokenistic and reaffirming of existing power structures (Gallacher and Gallagher, 2008; Lundy, 2007, 2018; UNMGCY, 2017; Warming, 2011). Josefsson and Wall (2020) noted a tendency for children to be represented in global policy-making through passive processes of hand-picked representation and tokenism. They urge global policymakers to create permanent and accountable structures to ensure that the fullest diversity of children's concerns are included. Children rarely have a guaranteed seat at the table. Policymakers should seek out diverse age perspectives so as to challenge their own assumptions and views.

Going forward, youth empowerment can occur through expanding their right to vote in local, regional and national elections. The inclusion of marginalised groups through their right to vote is a struggle that has a long history in the struggle for inclusion, diversity and justice and, as Josefsson and Wall (2020) pointed out, there are many sound philosophical arguments for expanding the franchise to young people (Beckman, 2006; Song, 2009; Wall, 2014). Josefsson and Wall (2020) also argued for a youth civil rights movement on the same global scale as for other historically disenfranchised groups and mutually supportive alliances between children and other marginalised groups.

Conclusion

Young people should be included as active contributors to future climate and environmental policymaking. Their ideas and interests should be incorporated into global climate governance. Children and young people are not helpless victims of climate change and environmental degradation. They are not passive, self-centred and uninterested in joining together to resolve environmental issues such as climate change. Their activism, however, challenges adult-centred concepts of the political and, certainly, adult-centred ideas of what constitutes acceptable civic behaviour for young people. The activists described here have defied these stereotypes and prescriptive boundaries in their variegated civic participation, activism and particularly in their dissent. The climate strikes brought global attention to the issues of climate change. They reminded the world of the importance of incorporating the voices of youth into deciding the policies and solutions for environmental issues. Their actions have supported attitude shifts, supporting the scientists and environmentalists who have been calling for urgent social transformation. Vitally, young climate activism, as evidenced by the climate strike movement, tends to be complex, justice-oriented and heterogeneous, calling for 'system change, not climate change'.

References

Alter, C., Haynes, S. and Worland, J. (2019) 'Person of the year: Greta Thunberg', *Time Magazine,* vol. 194, pp. 27–28. Available online: https://time.com/person-of-the-year-2019-greta-thunberg/.

Ambusaidi, A., Boyes, E., Stanisstreet, M. and Taylor, T. (2012) 'Omani students' views about global warming: Beliefs about actions and willingness to act', *International Research in Geographical and Environmental Education,* vol 21, no 1, pp. 21–39.

Amponsem, J., Doshi, D., Toledo, A.I.S., Schudel, L. and Delali-Kemeh, S. (2019) 'Adapt for our future: Youth and climate change adaptation'. A background paper commissioned by the Global Commission on Adaptation. Rotterdam and Washington, DC.

Austin, R. and Schill, M. (1991) 'Black, brown, poor & (and) poisoned: Minority grassroots environmentalism and the quest for eco-justice distributional consequences part II', *Kansas Journal of Law and Public Policy,* vol 1, pp. 69–82.

Bang, H. (2005) 'Among everyday makers and expert citizens', in Newman, J. (ed.), *Remaking Governance: Peoples, Politics and the Public Sphere.* Policy Press, Bristol.

BBC News (2020) 'Who is Greta Thunberg, the teenage climate change activist?', 28 January.

Beckman, L. (2006). 'Citizenship and voting rights: Should resident aliens vote?', Citizenship Studies, vol 10, no 2, pp. 153–165.

Boulianne, S., Lalancette, M. and Ilkiw, D. (2020) '"School Strike 4 Climate": Social media and the international youth protest on climate change', *Media and Communication,* vol 8, no 2, pp. 208–218.

Bowman, B. (2019) 'Imagining future worlds alongside young climate activists: A new framework for research', *Fennia – International Journal of Geography,* vol 197, no 2, pp. 295–305.

Bowman, B. (2020) '"They don't quite understand the importance of what we're doing today": The young people's climate strikes as subaltern activism', *Sustainable Earth,* vol 3, no 1, pp. 1-13.

Buckley, C.G. (2020) 'Climate change and the contemporary novel by Adeline Johns-Putra', *C21 Literature: Journal of 21st-Century Writings,* vol 8, no 1.

Checkoway, B. (2011) 'What is youth participation?' *Children and Youth Services Review,* vol 33, pp. 340–345.

Chhokar, K., Dua, S., Taylor, N., Boyes, E. and Stanisstreet, M. (2011) 'Indian secondary students views about global warming: Beliefs about the usefulness of actions and willingness to act', *International Journal of Science and Mathematics Education*, vol 9, no 5, pp. 1167–1188.

Costanza-Chock, S. (2012) *Youth and Social Movements: Key Lessons for Allies*. Berkman Center Research Publication, Cambridge, MA.

Council on Environmental Health (2015) 'Global climate change and children's health', *Pediatrics*, vol 136, no 5, pp. 992–997.

Crick, B. (1999) 'The presuppositions of citizenship education', *Journal of the Philosophy of Education*, vol 33, no 3, pp. 337–352.

Cullis-Suzuki, S. (1992) 'Speech at UN Conference on Environment and Development', available online: www.americanrhetoric.com/speeches/severnsuzukiunearthsummit.htm.

Curnow, J. and Helferty, A. (2018) 'Contradictions of solidarity'. *Environment and Society*, vol 9, no 1, pp. 145–163.

Currie, J. and Deschênes, O. (2016) 'Children and climate change: Introducing the issue', *The Future of Children*, vol 26, no 1, pp. 3–9.

Declaration on Children, Youth, and Climate Action (2019). Available online: https://www.childrenvironment.org/declaration-children-youth-climate-action

Eskenazi, B., Etzel, R.A., Sripada, K., Cairns, M.R., Hertz-Picciotto, I., Kordas, K., Suárez-López, J.R. (2020) 'The International Society for Children's Health and the Environment commits to reduce its carbon footprint to safeguard children's health', *Environmental Health Perspectives,* vol 128, p. 014501.

Evensen, D. (2019) 'The rhetorical limitations of the #FridaysForFuture movement', *Nature Climate Change*, vol 9, no 6, pp. 428–430.

Fridays for Future (2019) *Lausanne Climate Declaration: Official Version* (Online). SMILE for Future, Lausanne. https://drive.google.com/file/d/1Nu8i3BoX7jrdZVeKPQShRycI8j6hvwC0/view.

Gallacher, L.-A. and Gallagher, M. (2008) 'Methodological immaturity in childhood research? Thinking through "participatory methods"', *Childhood,* vol 15, no 4, pp. 499–516.

Gobby, J. (2019) 'Climate Justice Montreal: Who we are and what we do', in Perkins, P.E. (ed.), *Local Activism for Global Climate Justice: The Great Lakes Watershed*. Routledge, Abingdon.

Guardian (2019) 'Climate strike named 2019 word of the year by Collins Dictionary', 7 November.

Hagedorn, G. et al. (2019) 'Concerns of young protesters are justified', *Science*, vol 364, pp. 139–140.

Hamilton Faris, J. (2019) 'Sisters of ocean and ice: On the hydro-feminism of Kathy Jetñil-Kijiner and Aka Niviâna's rise: From one island to another', *Shima: The International Journal of Research into Island Cultures*, vol 13, no 2, pp. 1-24.

Han, H. and Ahn, S.W. (2020) 'Youth mobilization to stop global climate change: Narratives and impact', *Sustainability*, vol 12, no 4127, pp. 2–23.

Harris, A., Wyn, J. and Younes, S. (2010) 'Beyond apathetic or activist youth: "Ordinary" young people and contemporary forms of participation', *YOUNG*, vol 18, no 1, pp. 9–32.

Haste, H., Bermudez, A. and Carretero, M. (2017) 'Culture and civic competence: Widening the scope of the civic domain', in García-Cabrero, B. and Sandoval-Hernández, A. (eds.), *Civics and Citizenship: Theoretical Models and Experiences in Latin America*. Sense, Rotterdam.

Haynes, J. (1999) 'Power, politics and environmental movements in the Third World', *Environmental Politics*, vol 8, no 1, pp. 222–242.

Josefsson, J. and Wall, J. (2020) 'Empowered inclusion: Theorizing global justice for children and youth', *Globalizations*, vol 17, no 6, pp. 1043–1060.

Lee, K., Gjersoe, N., O'Neill, S. and Barnett, J. (2020) 'Youth perceptions of climate change: A narrative synthesis' *WIREs Climate Change*, vol 11, no 3, pp. 1-24

Lundy, L. (2007) '"Voice" is not enough: Conceptualising Article 12 of the United Nations Convention on the Rights of the Child', *British Educational Research Journal*, vol 33, no 6, pp. 927–942.

Lundy, L. (2018) 'In defence of tokenism? Implementing children's right to participate in collective decision-making', *Childhood*, vol 25, no 3, pp. 340–354.

Majeed, H. and Lee, J. (2017) 'The impact of climate change on youth depression and mental health', *Lancet Planetary Health*, vol 1, pp. e94–e95.

Malandrakis, G., Boyes, E. and Stanisstreet, M. (2011) 'Global warming: Greek students' belief in the usefulness of pro-environmental actions and their intention to take action', *International Journal of Environmental Studies*, vol 68, no 6, pp. 947–963.

Milman, O. (2019) 'Greta Thunberg condemns world leaders in emotional speech at UN', *Guardian,* 23 September.

Munawar, A. (2019) 'Social media reactions: Greta Thunberg, UN Climate Action Summit 2019', *Meltwater*, 1 October.

Myers, V. (2019) 'Global climate strike demonstrates mass movement to address climate change', American Federation of Teachers. Available online: www.aft.org/news/global-climate-strikedemonstrates-mass-movement-address-climate-change.

O'Brien, K., Selboe, E. and Hayward, B.M. (2018) 'Exploring youth activism on climate change: Dutiful, disruptive, and dangerous dissent', *Ecology and Society*, vol 23, no 3, pp. 1-13.

O'Toole, T., Lister, M., Marsh, D., Jones, S. and McDonagh, A. (2003) 'Tuning out or left out? Participation and non-participation among young people', *Contemporary Politics,* vol 9, no 1, pp. 45–61.

Ojala, M. (2012a) 'Hope and climate change: The importance of hope for environmental engagement among young people', *Environmental Education Research*, vol 18, no 5, pp. 625–642.

Ojala, M. (2012b) 'Regulating worry, promoting hope: How do children, adolescents, and young adults cope with climate change?' *International Journal of Environmental and Science Education*, vol 7, no 4, pp. 537–561.

Olson, J. (2016) 'Youth and climate change: An advocate's argument for holding the US government's Feet to the Fire', *Bulletin of the Atomic Scientist*, vol 72, no 2, pp. 79–84.

Otto, S. and Pensini, P. (2017) 'Nature-based environmental education of children: Environmental knowledge and connectedness to nature, together, are related to ecological behaviour', *Global Environmental Change*, vol 47, pp. 83–94.

Ouellett, M.L. (1996) 'Systemic pathways for social transformation: School change, multicultural organization development, multicultural education, and LGBT youth', *Gay Lesbian Bisexual Identity*, vol 1, pp. 273–294.

Our Children's Trust (n.d.) *Juliana v. U.S.* Available online: www.ourchildrenstrust.org/juliana-v-us.

Parker, L. (2019) 'Kids suing governments about climate: It's a global trend', *National Geographic*, 26 June. Available online: www.nationalgeographic.com/environment/2019/06/kids-suing-governmentsabout-climate-growing-trend/.

Pickard S. (2018) 'Momentum and the movementist "Corbynistas": Young people regenerating the Labour Party in Britain', in Pickard S. and Bessant J. (eds.), *Young People Re-Generating Politics in Times of Crises*. Palgrave Macmillan, London.

Pickard, S. (2019) *Politics, Protest and Young People: Political Participation and Dissent in 21st Century Britain*. Palgrave Macmillan, London.

Pulido, L. (1996) *Environmentalism and Economic Justice: Two Chicano Struggles in the Southwest.* University of Arizona Press, Tucson, AZ.

Pulido, L. (1997) 'Community, place and identity', in Jones, J.P., Nast, H.J. and Roberts, S.M. (eds.), *Thresholds in Feminist Geography: Difference, Methodology, Representation*. Rowman and Littlefield, Lanham, MD.

Sanson, A.V., Van Hoorn, J. and Burke, S.E. (2019) 'Responding to the impacts of the climate crisis on children and youth', *Child Development Perspectives*, vol 13, pp. 201–207.

Scientists for Future (2019) 'The concerns of the young protesters are justified. A statement by Scientists for Future concerning the protests for more climate protection', GAIA, vol 28, no 2, pp. 79 – 87.

Skamp, K., Boyes, E. and Stannistreet, M. (2009) 'Global warming responses at the primary secondary interface: Students' beliefs and willingness to act', *Australian Journal of Environmental Education*, vol 25, pp. 15–30.

Sloam, J. (2020) 'Young Londoners, sustainability and everyday politics: The framing of environmental issues in a global city', *Sustainable Earth,* vol 3, no 1, pp. 1-7.

Song, S. (2009). 'Democracy and noncitizen voting rights', Citizenship Studies, vol 13, no 6, pp. 607–620.

Spivak, G.C. (1988) 'Can the subaltern speak?', in Nelson, C. and Grossberg, L. (eds.), *Marxism and the Interpretation of Culture*. University of Illinois Press, Champaign, IL.

Spivak, G.C. (1999) *A critique of Postcolonial Reason: Toward a History of the Vanishing Present.* Harvard University Press, Cambridge, MA.

Taylor, M., Watts, J. and Bartlett, J. (2019) 'Climate crisis: 6 million people join latest wave of global protests', *The Guardian*, 27 September.

Thomas, A., Cretney, R. and Hayward, B. (2019) 'Student strike 4 climate: Justice, emergency, and citizenship', *New Zealand Geographic*, vol 75, pp. 96–100.

Thunberg, G. (2019a) 'A speech at the World Economic Forum', 22 January. Available online: www. fridaysforfuture.org/greta-speeches#greta_speech_jan22_2019.

Thunberg, G. (2019b) 'Our house is falling apart, and we are rapidly running out of time. A Speech to EU Parliament. Strasbourg, Germany', 16 April. https://speakola.com/ideas/gretathunberg-speech-to-eu-parliament-2019

Thunberg, G. (2019c) 'Greta Thunberg to world leaders: "How dare you – you have stolen my dreams and my childhood" – video', *The Guardian,* September 23. www.theguardian.com/environment/video/2019/sep/23/greta-thunberg-to-world-leaders-how-dare-you-you-have-stolen-my-dreams-and-my-childhood-video.

UN Development Program (2015) 'Fast facts: Youth and climate change', November 2015. Available online: https://reliefweb.int/sites/reliefweb.int/files/resources/FF-Youth-Engagement-Climate%20Change_.

UN (2020) 'Climate action: Youth summit', Available online: www.un.org/en/climatechange/youth-summit. shtml

UN News (2019) 'At UN, youth activists press for bold action on climate emergency, vow to hold leaders accountable at the ballot box', 21 September. https://news.un.org/en/story/2019/09/ 1046962

UNMGCY/ United Nations Major Group for Children and Youth (2017) 'Principles and barriers for meaningful youth participation'. https://static1.squarespace.com/static/5b2586e41aef1d89f00c60a9/t/5b2bc1fb2b6a286265e3a850/1529594364048/UN+MGCY-+Principles+and+Barriers+for+Meaningful+Youth+Engagement.pdf.

Vromen, A., Loader, B.D. and Xenos, M.A. (2015) 'Beyond lifestyle politics in a time of crisis?: Comparing young peoples' issue agendas and views on inequality', *Policy Studies*, vol 36, no 6, pp. 532–549.

Wahlström, M., Kocyba, P. de Vydt, M. and de Moor, J. (eds.) (2019) 'Protest for a future: Composition, mobilization and motives of the participants in Fridays for Future climate protests on 15 March, 2019 in 13 European cities'. https://protestinstitut.eu/wp-content/uploads/2019/07/20190709_Protest-for-a-future_GCS-Descriptive-Report.pdf.

Wall, J. (2014). 'Why children and youth should have the right to vote: An argument for proxy-claim suffrage', Children, Youth and Environments, vol 24, no 1, pp. 108–123.

Warming, H. (2011) 'Children's participation and citizenship in a global age: Empowerment, tokenism or discriminatory disciplining?', *Social Work and Society*. vol 9, no 1, pp. 119–134.

Warren, M. (2019). 'Thousands of scientists are backing the kids striking for climate change'. Available online: www.nature.com/articles/d41586-019-00861-z.

Wood, B.E. (2014) 'Researching the everyday: Young people's experiences and expressions of citizenship', *International Journal of Qualitative Studies in Education*, vol 27, no 2, pp. 214–232.

YOUNGO (Children and Youth Constituency to the United Nations Framework Convention on Climate Change) (2019). http://www.youngo.uno/

Young, L. (2018) *Rise: How Jeremy Corbyn Inspired the Young to Create a New Socialism*. Simon & Schuster, London.

10
POLICIES AND CHANGE

Karen Bell

Introduction

This book aimed to build bridges between those focussed on environmental issues and those engaged with equalities issues, whether as activists, policymakers, planners or academics. While some will be working simultaneously for environmentalism and diversity, there continues to be tensions and misunderstandings between the two areas of action, policy, planning or study. The chapter authors have articulated some of the problematic dynamics, power asymmetries and contradictions, making suggestions for how to resolve or overcome them. Building on their work, this chapter identifies some of the common themes discussed in the book and suggests some general policies for developing a unified, intersectional environmentalism. The first part is structured around three cross-cutting themes: misrecognition, power differentials and economic inequality. The second, and final, part makes suggestions on how to overcome the barriers to inclusion and diversity that are undermining an effective transition to sustainability. The strategies and policies suggested are the following: improving participatory practice, supporting joint social and environmental justice campaigns, working for the reduction of inequalities in society, choosing environmental solutions which meet the needs of marginalised groups and changing the political-economic system towards one that does not require dehumanisation and division.

As discussed in Chapter 1, when discussing the eight socio-economic groups addressed in this book collectively, we use one of the following terms: 'marginalised/ disadvantaged/equalities/oppressed' and 'people/groups/communities'. While not exactly applying in every case, they are used here as short-hand terms. The three cross-cutting themes that span the groups will now be discussed in turn.

Misrecognition

Throughout this book, there are stories and studies that point to the pervasive 'mis-recognition' of disadvantaged groups in mainstream environmentalism. For example, Clara Greed in Chapter 3 argues that environmental planning has demonstrated a lack of awareness of women's different experiences of life, both inside and outside the home. Similarly, Gary Haq in Chapter 8 writes about how older people are often invisible or misrecognised. 'Recognition' in environmentalism is about whose participation and perspectives are given weight, and what kind of experiences and knowledges about the environment are recognised (or not) in environmental discussion and decision-making. Many of the issues raised in this book are questions of recognition. There include stories and studies indicating the dismissal of some voices or denying us a seat at the table where decisions that impact our lives are being made. Different ideas of 'the environment' or environmentalism' can be part of this struggle for recognition. An important example of this is, as Silpa Satheesh emphasises in Chapter 4, the absence of the environmental struggles of the Global South in the mainstream literature on environmentalism. Similarly, in Chapter 7, Roger Griffith and Gnisha Bevan highlight how the environmental activism of Black, Asian and Minority Ethnic people is often overlooked. These omissions have resulted in conceptions of environmentalism as White, middle-class and rooted in the Global North.

It is also important to consider whose environmental knowledge and experiences are considered to be relevant in any environmental decision-making scenario. Often, only certain kinds of environmental knowledge and concern are recognised as legitimate (Haluza-Delay et al., 2009; Schlosberg, 2003, 2004, 2007). Sometimes the exclusion of views and information from marginalised groups is justified on the basis of the group's manner of engagement. Yet, as some scholars have argued (e.g. Peterson and Lupton, 1996), individuals should not be expected to conform to a 'participatory ideal'. Different life experiences will bring diverse forms of communication and knowledge. For example, with a history of oppression, some members of marginalised groups can lack the confidence or experience to present themselves orally in a public meeting. Their contributions can be dismissed if they are not presented according to the unwritten rules and expectations of the dominant group. When environmental problems can only be framed in very 'scientific', 'technical' or 'intellectual' ways, the debates can become inaccessible for some (Gibson-Wood et al., 2012).

Misrecognition is connected to 'discrimination' and both are often underpinned by stereotyping. Negative stereotypical generalisations can be used to justify exclusion and inequity in terms of the supposed shortcomings of the targets. Yet, throughout this book, the authors stress that the groups they write about are not homogeneous, except to the extent that they can experience prejudice. Even so, stereotypes persist because the 'truths' and norms of society tend to be determined by the wealthy and powerful who play a key role in the distribution of knowledge. This reinforces the idea that the situation of the marginalised is justifiable, and that their low incomes and low status are the result of their own deficiencies. As studies by the psychologist Paul Piff and his colleagues demonstrate, wealthier people tend to believe that they deserve greater rewards, even when it is obvious to all that the game is rigged in their favour (Piff et al., 2012).

Misrecognition and discrimination can occur without deliberate intent, through habitual historic, cultural and institutionalised practices. In a situation of crisis, where policy may be developed in haste without consideration of its impacts on marginalised groups, stereotypes may become further embedded. Gary Haq discusses in Chapter 8, for example, how the coronavirus (COVID-19) pandemic may have reinforced ageist views of older people as a homogeneous group associated with death, dependency and high vulnerability. It is important, then, not to fall back on stereotypes in the urgent search for policy solutions to our environmental crises. It is also important to avoid misrecognition in developing these solutions, as occurs, for example, when environmental policies impose additional burdens on marginalised and low-income groups. It is evident then that misrecognition is underpinned by power differentials, the topic of the next section.

Power differentials

When mainstream environmental groups' 'open door' policies or attempts at outreach fail, some might imagine that this is because marginalised groups are 'apathetic' and 'passive' with regard to this issue. It is true that over a lifetime of not being listened to or taken seriously, marginalised people can stop trying to be heard and cease trying to change their lives or the situation in their communities. Psychological studies on 'learned helplessness' show that when humans and many other animals endure repeatedly painful or otherwise aversive stimuli which they are unable to escape or avoid, they learn to stop trying to avoid the situation, even when the opportunity to leave presents itself (e.g. Seligman, 1972). Those who have lived in deprived communities often comment on the hopelessness that is so commonly encountered in these areas. For example, rapper and social commentator, Darren McGarvey (2017, p. 48), describes in *Poverty Safari* how there is 'a pervasive belief that things will never change' on the working-class estate that he was brought up on. He notes that this is partly a result of the continual barriers that people come up against when they try to change anything.

However, even though some may have fallen into despair, in my experience, there are always at least some people from these marginalised groups that are willing to work for environmentalism. But, when they attempt to engage with environmental organisations, they sometimes find that they are less able to influence decisions than their non-marginalised counterparts. Their engagement can, therefore, be frustrating and even more disempowering. Hence, Chapter 9 asks, with regard to young people's climate activism, and drawing on Spivak (1988), 'how can power be challenged by subaltern groups?' To do this, as this book highlights, we must begin with considering the difference between diversity and inclusion. It is important to go beyond bringing in marginalised groups, to ensure that they are met with inclusive spaces, respect and the ability to exert appropriate influence. As Velicu and Barca (2020, (p. 266) note, 'once at the table of negotiation (often celebrated as an environmental justice victory), people are further exposed to violence, patronized, intimidated, manipulated, coopted, or simply disregarded as uninformed and unreasonable'. This is particularly well illustrated in Chapter 2, where Harriet Larrington-Spencer and colleagues discuss the dismissal of disabled people as 'kill-joys' when they articulate their access requirements. These incidents need to be prevented so that marginalised and disadvantaged groups can be better involved in scrutinising and analysing scientific information, setting standards that reflect their needs and experience, and deciding on the appropriate interventions and solutions to address environmental problems. Their voices are important for effectively addressing the environmental crises in ways that work for all.

It is particularly difficult for disadvantaged people to take on the powerful corporate lobbies that shape environmental policies (see Faber 2008; Magdoff and Foster 2011). These economic interests can undermine attempts to amplify the voices from below. This perhaps explains why most of the ecological crises are continuing to escalate (Dempsey et al., 2011; Steffen et al., 2015; Dauvergne, 2016). For example, the Civil Society Reflection Group on the 2030 Agenda for Sustainable Development (2018, p. 11) stated that most governments have failed to create concrete policies to achieve the goals and, in some countries, changes are occurring in the opposite direction. Yet, the environmental movement has sometimes not been willing to challenge the excesses and irrationality of powerful economic interests. Hence, Dauvergne (2016) has argued that we have an 'environmentalism of the rich', focusing on eco-business and eco-consumption. The solutions offered have frequently been individualised, advocating lifestyle changes which, in some cases, can create more environmental problems than they solve, such as electric cars (see Hawkins et al. 2012; Monbiot, 2020). Disproportionately emphasising individual behaviour and lifestyle solutions can alienate marginalised groups who often have fewer options. As this book highlights throughout, it is important to consider social conditions, rather than just individual decisions, and to understand how these constrain personal choices.

When focusing on 'educating' ordinary citizens, mainstream environmentalists have not been paying attention to the changes that the state or corporations need to make. This is often where the greatest environmental degradation is generated. For example, Sanders' (2009) work on the military impacts on the environment indicates that even if all of us made all the recommended green lifestyle changes, we would still be on a trajectory to environmental and climate disaster because of the damage from the military alone. Yet this topic is rarely mentioned in most environmental debates.

Economic inequality

Globally, inequalities within individual nations are generally increasing (Hickel, 2018). This is undermining sustainability efforts, in terms of inclusion in environmentalism, as well as environmental degradation. For example, as mentioned throughout the book, many barriers to environmentalism occur as a result of the relative financial constraints of disadvantaged people. Being an environmental activist has direct and indirect financial costs which constrain some disadvantaged groups from participating. It is important to recognise how material inequalities are hugely influential in providing or denying opportunities for participation in civic life (Wakefield and Poland, 2005). Free time has been eroded even further in recent years as people are more likely to be employed on casual contracts where it is difficult for them to plan ahead. It is also the case that if we did not have the wider inequalities in society that produce hierarchies, and the associated parallel lives, we would be less segregated and find it easier to organise campaigns across diverse groups.

We also know that inequality contributes to environmental destruction. Studies have shown a link between inequality and worse environmental outcomes in terms of levels of consumption (Wilkinson and Pickett, 2010), industrial pollutant waste (e.g. Jun et al., 2011), air pollution (e.g. Torras and Boyce, 1998), biodiversity (e.g. Pandit and Laband, 2009), energy consumption (e.g. Baek and Gweisah, 2013), the number of green technology patents (e.g. Vona and Patriarca, 2011) and environmentally friendly behaviour (e.g. Dorling, 2010, 2011). For example, the populations of the most unequal of the High Income Countries tend to consume more (Dorling, 2011, 2017; Oxfam, 2016). Dorling (2011, 2017) has described how higher levels of inequality in a country tend to correlate with a larger per capita ecological footprint; greater consumption of meat, water and flights; and the production of more waste. This may be a result of 'emulative consumption' (Veblen, 1994 [1899]), where, in more unequal societies, displays of material wealth are used as a means to gain or maintain respect. The levels of consumption often reflect the need to maintain status and gain acceptance. Studies have shown that, as a social species, the need for the approval of our peers and avoidance of rejection is very strong (e.g. Weir, 2012). The consequent consumer culture causes further psychological problems by encouraging competitiveness, overconsumption and indebtedness, likely contributing to the rising incidence of depression, anxiety and addictions globally (Rehm and Shield, 2019).

Wilkinson and Pickett (2010, 2019) have argued that inequality damages the whole society, not just the least well-off and marginalised. Many studies have shown a correlation between inequality and a range of social problems, including crime (Gallan, 2018), unwanted pregnancy (e.g. Kawachi et al., 1997), less trust in others (e.g. Fiske et al., 2012; Fritsche et al., 2017), poor health, unhappiness (e.g. Oishi et al., 2011), mental illness (Royal College of Psychiatrists, 2010), drug addiction, obesity, loss of community life, imprisonment, childhood disadvantage, increased personal debt (Wilkinson and Pickett, 2010), child abuse (Eckenrode et al., 2014) and bullying (Due et al., 2005). In the United Kingdom (UK), one of the most unequal of the High Income Countries, one in four adults has been diagnosed with a mental illness, and 4 million people take antidepressants every year. The Royal College of Psychiatrists states that:

> Inequality is a major determinant of mental illness: the greater the level of inequality, the worse the health outcomes. Children from the poorest

households have a three-fold greater risk of mental ill health than children from the richest households. Mental illness is consistently associated with deprivation, low income, unemployment, poor education, poorer physical health and increased health-risk behaviour.

(Royal College of Psychiatrists, 2010, p. 18)

Moreover, although some might argue that the chance of becoming rich is a great motivator for human endeavour, excess wealth has huge ecological consequences through increased ability to consume. Above a certain basic level, acquiring more wealth does not even increase happiness (Easterlin et al., 2010).

Inequality is not only harmful but also unnecessary. In 2013, Oxfam reported that the world's 100 richest individuals earned enough in the prior year to end extreme poverty, worldwide, four times over (Oxfam, 2013). However, inequality is often justified as inevitable or even beneficial (Hadler, 2005). Despite pockets of resistance and organisation against specific forms of inequality, such as the Occupy movement, Black Lives Matter and the Equality Trust, in general, greater economic equality is not something that is yet being widely demanded. We have tended to focus instead on social mobility, even while this continues to be achievable for relatively few. The UK rate of social mobility is currently below average internationally and is one of the lowest in Europe (Social Mobility Commission, 2017).

Social mobility assumes that a hierarchical society is inevitable and the problem is only how to best climb the ladder. Yet social mobility just allows a few individuals from marginalised and oppressed groups to become wealthier and more powerful. Although they might try to smooth the way for others from a similar background, this will be a very slow transition that will take generations. If we are all going to live within planetary limits, we need to share better and this will require a significant redistribution of wealth, income and power. There has been too much focus on 'equal opportunities' and not enough on 'equality'. We need more than opportunities – we need outcomes. Campaigning for greater equality, alongside calls for sustainability, would strengthen both the social justice and the environmental movements.

Policies going forward

Improve participatory practice

As discussed in Chapter 1 and throughout the book, diversity and inclusion leads to greater success and creativity in organisations. Often the best way to ensure diversity and inclusion is to be led by the marginalised and disadvantaged groups themselves. Leadership by members of these groups is key to genuine participation as Arnstein (1969) highlighted in her 'ladder of participation'. As she emphasised, there is,

> …[A] critical difference between going through the empty ritual of participation and having the real power needed to affect the outcome of the process…participation without redistribution of power is an empty and

frustrating process for the powerless. It allows the powerholders to claim that all sides were considered, but makes it possible for only some of those sides to benefit. It maintains the status quo.

(Arnstein, 1969, p. 216)

One way to ensure the leadership of marginalised groups is to consider appointing members of those groups to paid roles. As Roger Griffith and Gnisha Bevan highlight in Chapter 7, a key issue in terms of properly valuing marginalised groups is that they are able to access any paid employment opportunities that arise. It is difficult for members of these groups to offer their time voluntarily due to their economic disadvantage (needing to work long hours or undertake multiple jobs to make ends meet on their low wages). A positive step has been the attempts to target marginalised and oppressed groups for green jobs. In the United States, for example, a report on how to ensure green jobs benefit under-represented groups states, '[t]he number of high-quality jobs that are created and filled should be maximized and incentivized, with an aim to distribute them proportionally by race, gender and income level' (Liu and Keleher, 2009, p. 14). However, these initiatives are still rare.

The agenda needs to be set by the marginalised groups, themselves, and this may require the relatively privileged to relinquish some of their power. Yet often, while keen to appear or be more diverse, the members of some projects and institutions may have no real interest in fully engaging with the different worldview that the marginalised member brings. A great deal of diversity work focuses on getting different marginalised groups involved in projects and institutions which have been designed by the groups that have oppressed them. This does not mean that diversity and inclusion is always just a box ticking exercise. Often there are good intentions. However, all organisations have their own cultures, and social pressure, albeit unconscious, fosters conformity with that culture. This means that the organisations and institutions need to be constantly mindful of the needs of the diverse members. These members bring a different perspective and may need space and time to develop the trust needed to allow them to express their views. Even if these views are unusual, the mainstream organisation or institution needs to work hard to see how their ideas could be incorporated and actioned, even if that would mean quite a radical change of direction from what has gone before.

Throughout this book, there is guidance on how to improve participatory practice. For example, in Chapter 7, Roger Griffith and Gnisha Bevan discuss the many ways to improve participatory practice for Black, Asian and Minority Ethnic people in environmental organisations. They offer specific guidance as to how to tackle gaps in environmental and civic leadership, and that can be modelled nationally and internationally.

Support joint social and environmental justice campaigns

In Chapter 6, Emma Foster discusses Gaard's (1997, p. 114) idea that the conservative right has tried to group environmentalist and equalities groups together in the hope of seeing their 'collective annihilation'. Yet, ironically, these groups remain disunited and even antagonistic to one another. Woven into this book are many examples of the lack of meaningful collaboration between equalities groups and environmentalists (as well as some positive initiatives). Mainstream environmental organisations have tended to start with their own priorities when they attempt to engage with marginalised communities instead of joining in with, and broadening, struggles for social justice. Hence, those employed in environmentally harmful jobs have sometimes found environmentalism to be a threat, not only to their jobs, but also to their identities (Barca and Leonardi, 2016). Environmentalists can then be seen to be part of a privileged group that can afford to make choices about their consumption and production that others cannot.

There is one particular issue that intersects equalities, social justice and environmentalism and that is health. Yet, as Craig Bennett, the former chief executive officer (CEO) of Friends of the Earth England and Wales, noted, this has been a:

> …[B]ig blind spot… It is an area where we haven't really made the connections we should have done over the last few decades. We've got really strong and established environment and health sectors and actually not that much collaboration between them at the moment. That is extraordinary.
>
> *(Bennett, 2015, n.p.)*

This is now beginning to change, for example, with more information on how pollution has impacted our health and how this may underpin some of the COVID-19 mortality patterns (e.g. Espejo et al., 2020). It is important that bringing discussions of health into environmentalism is not done using disablist terms, as is cautioned in Chapter 2. If these discourses can be developed with disabled people, environmental movements and marginalised people could be stronger allies when focused on the question of health. Environmental justice campaigners have frequently focused on health issues but struggle to convince others of the health impacts of environmental damage. Overall, though, climate justice, environmental justice and energy justice studies and activism are helping to break down this separation. Environmental issues impact on the health and well-being of all of us. Therefore, if environmentalists can approach disadvantaged groups to work on their own agendas around those issues, they may find their messages more likely to be embraced.

Even when the immediate issue that the equalities group is facing does not seem to connect with environmentalism, it is still important for environmentalists to proactively support their struggle. This can build the relationships and trust which is essential for solidarity. There are always mutual campaigns to be found. For example, as Silpa Satheesh highlights in Chapter 4 on the Global South, the struggle against environmental injustice is often a struggle to secure economic livelihoods. Emma Foster in Chapter 6 calls for movements to recognise the mutually reinforcing character of the oppressions they are facing and to combine their efforts into a broader anti-oppression movement. In particular, she emphasises that queer and environmental campaigns need to acknowledge that violence towards queers is inextricably linked to violence against nature, people of colour, women, immigrants, indigenous people and other-than-human animals.

It is important also to recognise that change has to be intersectional. Feminists have emphasised the need for alliances to form with multiple dispossessed groups (Burns, 2006; Feminist Alliance of Rights, 2019). If equalities groups are supporting each other, then they can become a major force within environmentalism. Gary Haq in Chapter 8 highlights the unhelpfulness of some young people seeking to 'blame' older people for the climate crisis. Marginalised groups should not be attacked in the course of making arguments for environmentalism, even while it is important to discuss the pressure and constraints on us that draw us into behaviours that can contribute to environmental degradation.

Work for the reduction of inequalities in society

Unmet needs in the midst of excess on a finite planet call for policies that will redistribute income and wealth. As Gough (2017, p. 179) argues, '[w]hen the cake shrinks, its distribution becomes critical'. One step towards this is to join and support labour and community unions. A number of studies have indicated that unionisation has a significant equalising effect on national income distribution, with less inequality across society where there is higher union membership (Dabla-Norris et al., 2015; Dromey, 2018; Farber et al., 2018).

Another step is to introduce eco-social policies, such as those advocated by Kate Raworth and Ian Gough. Kate Raworth's (2017) doughnut-shaped diagram represents both the productive limits of the earth's systems, the ceiling that should not be exceeded, and the minimum foundation for human survival and well-being. This illustrates how we should develop policies which set a lower threshold of well-being, below which no one should fall, and an upper threshold of environmental limits that should not be transgressed. We need an adequate minimum but also a maximum to avoid overstepping planetary boundaries. This could include a cap on excessive incomes and wealth. With similar aims, Gough (2017) advocates widening social consumption, that is, making provision for more public goods, as this would reduce opportunities for people to compare consumption, one of the drivers of excess. Greater social consumption is more ecologically efficient than private consumption, partly as a result of a better allocation of resources. Similarly, Monbiot argues that we need 'private sufficiency and public luxury' so that we could share the goods and services that we do not need to individually own, that is, free-at-the-point-of-use swimming pools, parks, playgrounds, sports centres, galleries, allotments and public transport (Monbiot, 2017a, n.p.). These ideas are also reflected in the call for Universal Basic Services from the Institute for Global Prosperity (IGP) (2017). The IGP recommended that, in the UK, we need to build more social housing units, offered for free to those in most need; supply free bus passes to the entire population; develop a food service for those who experience food insecurity; and provide basic phone and Internet services, as well as free TV licences. The idea to provide an ecological universal basic income (UBI), that is, a regular payment from the state to every citizen on ecological grounds, is another possible eco-social policy. If managed well, an ecological UBI could run alongside social programmes, services and 'green jobs'. There are many socially useful jobs that need doing but are either not done or not paid for, because they are not profit-making. A UBI payment could pay people to do those important jobs.

On an international scale, 'contraction and convergence' of greenhouse gas (GHG) emissions across nations has also been suggested as a way of addressing climate justice. Every country would bring its GHG emissions, per capita, to a level which is equal for all countries, leading to a contraction for some countries, a growth for others and an overall convergence (Meyer, 2000). According to this strategy, in the Low- and Middle-Income Countries, economies would grow and redistribute wealth until the basic needs of all are met, at which point these countries would also stabilise their growth (e.g. Lawn and Clarke, 2010).

While these policies would require greater state control and investment, in recent years there have been growing calls for greater levels of state intervention, including nationalisation of utilities (Legatum Institute and Populus, 2017). For example, as I write, there is currently a public outcry in the UK that the government refuses to extend food programmes for children into the school holidays during the COVID-19 pandemic (Chakelian, 2020). This indicates that, when people understand that we are facing a crisis, they no longer believe as much in selfish individualism. By implementing the above policies, we would ensure that everyone has support to lead a dignified, decent and pleasant life in the transition to sustainability.

Choose environmental solutions which meet the needs of marginalised groups

The transition to sustainability will require major transformation in work and life for many of us. Marginalised groups may be more likely to embrace the measures necessary to do this if they are not going to increase the problems in their, often already difficult, daily lives. Importantly, these measures should not lead to unemployment, underemployment, a drop in income, increased costs, or other threats to survival, well-being or identity. Instead of describing the transition to sustainability in terms of 'sacrifice' and job losses, we need to be talking about social protection for all and a life of well-being and respect within the limits of the planet. Some analysts and activist, particularly trade unions, have emphasised that the transition to sustainability is an epochal opportunity for improving the lives of the working-class and marginalised (ITUC, 2015; Rosenberg, 2017).

In general, as argued earlier, environmentalism has often been very individualistic, masking the more severe environmental harms caused by unsustainable agriculture, transport, housing and defence policies. This approach encourages an overemphasis on individual action, at the expense of systemic or regulatory action. To be more inclusive for marginalised groups, environmental solutions should simultaneously be social solutions and address structural harms. Here, I list some environmental policies that would do this.

Firstly, environmental policies should improve daily life. For example, provide cheaper and more reliable bus services, affordable and healthy food, and opportunities for human connection. The social movement literature indicates that it is easier to mobilise around issues when people can see how an issue affects their everyday lives or the lives of those they care about (e.g. McAdam et al., 1996). Marginalised people, who may be very preoccupied with discrimination, poverty and dealing with the structural barriers of society, will be more likely to find time for environmentalism if it is connected to their self-identified well-being or their other core values.

Secondly, connect the local and the global in environmental policies so as not to 'outsource' our pollution and other environmental problems to Low-Income Countries or communities. We should be careful to ensure that workers and communities cannot be played off against each other around the world in a race to the bottom of lowering environmental standards to attract business. If our governments do not facilitate this, it will require mutual international solidarity via labour unions and other civil society organisations to ensure the development of this solidarity.

Thirdly, instead of targeting the practices of individuals, focus on the regulating of the companies and governments as the main culprits of unsustainability. As Clara Greed argued in Chapter 3, environmentalism is often preoccupied with car use without considering why many people use cars in the first place, that is, the very real time and distance constraints they experience in their daily lives. Environmental policies need to take into account these constraints.

Fourthly, build coalitions between workers, environmentalists, and equalities and social justice organisations. The notion of 'Just Transition' initiated by trade unions has been a positive step forward in this respect (see, e.g., Stevis et al., 2020). An employment strategy based on a worker-led programme for socially useful and environmentally sound production is required. We can still find that some 'green jobs' are thought about within an economy of high productivity and high consumption, which runs contrary to the idea of planetary boundaries. But many care and repair jobs, which do not drive increased consumption, are needed to improve human well-being. However, these jobs are not created in the numbers required because they are not profitable. For example, more home care and support workers are needed to meet the access needs of disabled people. Yet adequate jobs are not created in this sector because of insufficient funding to employ workers (Brawn et al., 2013).

Although some of these policies would probably require government subsidies, we know that governments are willing to spend in a crisis, as has occurred during the COVID-19 pandemic. The environmental crisis is also serious and threatens human survival. Some of these subsidies could occur through reducing government subsidies in unhelpful areas of the economy, such as the arms industry.

In this book, beacons of good practice are discussed in terms of environmental solutions that meet the needs of marginalised groups. For example, Larrington-Spencer and colleagues in Chapter 2 discussed the UK Department for Transport (DfT, 2020) 'Gear Change' report which provides national guidance to local authorities on best practice for cycle infrastructure design. The report and associated guidance is particularly visionary in terms of disabled environmentalism and disabled cycling, in that disabled people's needs are centralised within the development of the cycle infrastructure design. It is important to learn from these examples of good practice.

Change the political-economic system

Although not discussed by all the authors in this book, some of the contributors, including myself, have critiqued the political-economic system as driving some of the problems we discuss. This indicates the need to transform this system, moving towards one that does not require dehumanisation and divide and rule. Inequalities are driven in part by capitalism where the accumulation of wealth among the owning and middle-classes has been at the cost of the environmental and social well-being of people and planet, more generally. A number of analysts, including myself, have been arguing for some time that capitalism is not compatible with sustainability (e.g. Bell, 2011, 2012, 2014, 2015, 2016, 2020; Klein, 2014; Monbiot, 2017b; Gough, 2017). Even if we are wrong about this, it is clear that in this situation of crisis and growing inequality, we really need to rethink why, how and for whom goods are produced. We need to be honest about who is causing environmental degradation and the pressures and constraints which drive this. As Silpa Satheesh discusses in Chapter 4, there are explicit Marxist orientations to many environmental struggles in the Global South, so this questioning of the dominant political-economy is already underway outside of mainstream environmentalism.

In most environmental discussions, the capitalist system is the elephant in the room. Capitalism is considered to be the only possible, if not the most advanced, way of organising society. Yet, as most of the authors in this book have argued, proper inclusion and diversity would be a challenge to capitalism. In most cases, we are asking for more than not being 'left behind' in the transition to sustainability. We are looking beyond the hegemonic eco-modernist, market-environmentalist, green-growth Western lifestyle. Such 'solutions' have been found to be unlikely to even reduce carbon adequately, let alone address the other multiple environmental and social crises we face (Hickel and Kallis, 2020). In many cases these solutions may even worsen them (Brand and Wissen, 2013; Goodman and Salleh, 2013). There is an increasing cry from socialist, working-class, ecofeminist, peasant and indigenous groups for counter-hegemonic visions of a post-carbon world, as documented in the alternative Rio + 20 declaration (Peoples Summit, 2012). Versions of this vision are also articulated by many social movements around the world, including the Climate Justice Alliance, Just Transition Alliance and Via Campesina (Morena and Krause, 2018) as well as some governments, such as the Bolivian government during MAS leadership (Bell, 2017).

Conclusion

The themes outlined above are multi-layered, with many factors reinforcing and interacting. It is evident that the marginalised and disadvantaged groups discussed in this book mostly care about the environment yet are put off some mainstream environmental movements and environmental policies by attitudes, behaviours and structures that exclude them or seem to threaten their well-being. These barriers are numerous and this book can only outline some. More research and thought is needed to understand the full extent of this exclusion and how to best overcome it. We have all been socialised into racism, classism, sexism, disablism, ageism, homophobia and other harmful mindsets. It will be an ongoing project to rid ourselves of these. This book has, hopefully, encouraged and supported environmentalists to take some further steps in this direction. It has also highlighted how the environment is important to those campaigning for diversity, inclusion and equality. There is an ongoing struggle to continue to build alliances for the mutual achievement of social and environmental justice. We need to continue to learn how to forge an alliance such that the causes of ecological sustainability, social justice and human emancipation are considered as one.

References

Arnstein, S. R. (1969) 'A Ladder of Citizen Participation', *Journal of the American Planning Association,* vol 35, pp216–224

Baek, J. and Gweisah, G. (2013) 'Does income inequality harm the environment?', *Energy Policy*, vol. 62, C, pp. 1434–1437.

Barca, S. and Leonardi. E. (2016) 'Working-class communities and ecology', in M. Shaw and M. Mayo (eds.), *Class, Inequality and Community Development*. Policy Press, Bristol.

Bell, K. (2011) 'Environmental justice: Lessons from Cuba', PhD thesis, University of Bristol.

Bell, K. (2012) 'Is socially-just degrowth compatible with capitalism?', *Degrowth Conference,* Montreal. www.degrowth.info/en/catalogue-entry/is-socially-just-degrowth-compatible-with-capitalism/.

Bell, K. (2014) *Achieving Environmental Justice*. Policy Press, Bristol.

Bell, K. (2015) 'Can the capitalist economic system deliver environmental justice?', *Environmental Research Letters,* vol 10, no 12, pp. 1–8.

Bell, K. (2016) 'Green economy or living well? Assessing divergent paradigms for equitable eco-social transition in South Korea and Bolivia', *Journal of Political Ecology*, vol 23, pp. 71–92.

Bell, K. (2017) '"Living well" as a path to social, ecological and economic sustainability', *Urban Planning*, vol 2, no 4, pp. 19–33.

Bell, K. (2020) *Working-Class Environmentalism: An Agenda for a Fair and Just Transition to Sustainability*. Palgrave, London.

Bennett, C. (2015) 'Green movement must escape its "white, middle-class ghetto", says Friends of the Earth chief Craig Bennett', *The Independent*, 4 July.

Brawn, E., Bush, M., Hawkings, C. and Trotter, R. (2013) *The Other Care Crisis: Making Social Care Funding Work for Disabled Adults in England*. Scope, Mencap, The National Autistic Society, Sense and Leonard Cheshire Disability, London.

Brand, U. and Wissen, M. (2013) 'Crisis and continuity of capitalist society-nature relationships: The imperial mode of living and the limits to environmental governance', *Review of International Political Economy*, vol 20, no 4, pp. 687–711.

Burns, L. (2006) *Feminist Alliances*. Rodopi, New York.

Chakelian, A. (2020) 'By refusing to extend free school meals, the government exposes its warped idea of poverty', *New Statesman*, 26 October.

Civil Society Reflection Group on the 2030 Agenca for Sustainable Development (2018) Spotlight on Sustainable Development 2018 Exploring new policy pathways www.2030spotlight.org/sites/default/files/spot2018/Spotlight_2018_web.pdf.

Dabla-Norris, E., Kochhar, K., Suphaphiphat, N., Ricka, F. and Tsounta, E. (2015) 'Causes and consequences of income inequality: A global perspective', International Monetary Fund. www.imf.org/external/pubs/ft/sdn/2015/sdn1513.pdf.

Dauvergne, P. (2016). *Environmentalism of the Rich*. MIT Press, Cambridge, MA.

Dempsey, N., Bramley, G., Power S. and Brown, C. (2011) 'The social dimension of sustainable development: Defining urban social sustainability', *Sustainable Development,* vol 19, pp. 289–300.

DfT (2020) 'Gear change. A bold vision for cycling and walking', Department for Transport, London.

Dorling, D. (2010) 'Social inequality and environmental justice', *Environmental Scientist*, vol 19, no 3, pp. 9–13.

Dorling, D. (2011) 'Is more equal more green?', *Green Party Conference.* University of Sheffield, 9–12 September.

Dorling, D. (2017) 'Is inequality bad for the environment?', *The Guardian,* 4 July.

Dromey, J. (2018) 'Power to the people: How stronger unions can deliver economic justice', IPPR, London.

Due, P., Holstein, B., Lynch. J., Diderichsen, F., Gabhain, S., Scheidt, P., et al. (2005) 'Bullying and symptoms among school-aged children: International comparative cross sectional study in 28 countries', *The European Journal of Public Health*, vol 15, no 2, pp. 128–132.

Easterlin, R.A., McVey, L.A., Switek, M., Sawangfa, O. and Zweig, J.S. (2010). 'The happiness-income paradox revisited', *Proceedings of the National Academy of Sciences,* vol 107, no 52, pp. 22463–22468.

Eckenrode, J., Smith, E.G., McCarthy, M.E., and Dineen, M. (2014) 'Income Inequality and child maltreatment in the United States', *Pediatrics*, vol 133, no 3, pp. 454–461.

Espejo, W., Celis, J.E., Chiang, G. and Bahamonde, P. (2020) 'Environment and COVID-19: Pollutants, impacts, dissemination, management and recommendations for facing future epidemic threats', *Science of the Total Environmen*, vol 747, 141314.

Faber, D. (2008) *Capitalizing on Environmental Injustice: The Polluter Industrial Complex in the Age of Globalization.* Rowman and Littlefield, Lanham, MD.

Farber, H.S., Herbst, D., Kuziemko, I. and Naidu, S. (2013) 'Unions and inequality over the twentieth century: New evidence from survey data', NBER Working Paper No. 24587, National Bureau of Economic Research.

Feminist Alliance of Rights (2019) 'Post-2015 women's coalition' http://femini stallianceforrights.org/wp-content/uploads/2017/01/Post2015-Womens-Coalition_ Response-to-Transforming-Our-World-Outcome-rev.pdf.

Fiske, S.T., Moya, M., Russell, A.M. and Bearns, C. (2012) 'The secret handshake: Trust in cross-class encounters', in S. T. Fiske and H. R. Markus (eds.), *Facing Social Class: How Societal Rank Influences Interaction.* Russell Sage, New York.

Fritsche, I., Moya, M., Bukowski, M., Jugert, P., de Lemus, S., Decker, O., Valor-Segura, I. and Navarro-Carrillo, G. (2017) 'The great recession and group-based control: Converting personal helplessness into social class ingroup trust and collective action', *Journal of Social Issues*, vol 73, no 1, pp. 117–137.

Gaard, G. (1997) 'Toward a queer ecofeminism', *Hypatia*, vol 12, no 1, pp. 114–137.

Gallan, P. (2018) 'Rising crime is symptom of inequality, says senior Met chief', *The Guardian,* 4 June.

Gibson-Wood, H., Wakefield, S., Vanderlinden, L., Bierefeld, M., Cole, D., Baxter, J. and Jermyn, L. (2012) '"A drop of water in the pool": Information and engagement of linguistic communities around a municipal pesticide bylaw to protect the public's health', *Critical Public Health,* vol, 22, no 3, pp. 341–353.

Goodman, J. and Salleh, A. (2013) 'The 'green economy', *Globalizations*, vol 10, no 3, pp. 411–424.

Gough, I. (2017) *Heat, Greed and Human Need. Climate Change, Capitalism and Sustainable Wellbeing*. Edward Elgar, Cheltenham.

Hadler, M. (2005) 'Why do people accept different income ratios? A multi-level comparison of thirty countries', *Acta Sociologica*, vol 48, no 2, pp. 131–154.

Haluza-Delay, R., O'Riley, P., Cole, P. and Agyeman, J. (2009) 'Speaking for ourselves, speaking together: Environmental justice in Canada', in J. Agyeman, P. Cole, R. Haluza-Delay and P. O'Riley (eds.), *Speaking for Ourselves: Environmental Justice in Canada*. UBC Press, Vancouver.

Hawkins, T.R., Singh, B., Majeau-Bettez, G. and Strømman, A.H. (2012). 'Comparative environonmental life cycle assessment of conventional and electric vehicles', *Journal of Industrial Ecology*, vol 17, no 1, pp. 158–160.

Hickel, J. and Kallis, G. (2020) 'Is green growth possible?', *New Political Economy*, vol 25, no 4, pp. 469–486.

Hickel, J. (2018) 'Is Inequality within countries getting better or worse?', www.jasonhickel. org/blog/2018/12/13/what-max-roser-gets-wrong-about-inequality.

Institute for Global Prosperity (2017) 'Social prosperity for the future: A proposal for Universal Basic Services', www.ucl.ac.uk/bartlett/igp/sites/bartlett/files/universal_ basic_services_-_the_institute_for_global_prosperity_.pdf.

International Trade Union Confederation (ITUC) (2015) 'Climate justice: There are no jobs on a dead planet', *Frontlines Briefing*, March 2015.

Jun, Y., Zhong-kui, Y. and Peng-fei, S. (2011) 'Income distribution, human capital and environonmental quality: empirical study in China', *Energy Procedia*, vol 5, pp. 1689–1696.

Kawachi, I., Kennedy, B.P., Lochner, K., and Prothrow-Stith, D. (1997) 'Social capital, income inequality, and mortality', *American Journal of Public Health*, vol 87, no 9, pp. 1491–1498.

Klein, N. (2014) *This Changes Everything: Capitalism vs. the Climate*. Simon & Schuster, New York.

Lawn, P. and Clarke, M. (2010) 'The end of economic growth? A contracting threshold hypothesis', *Ecological Economics,* vol 69, pp. 2213–2223.

Legatum Institute and Populus (2017) 'Jeremy Corbyn's nationalisation plans are music to ears of public', *The Guardian*, 1 October.

Liu, Y. and Keleher, T. (2009) *The Green Equity Toolkit: Standards and Strategies for Advancing Race, Gender and Economic Equity in the Green Economy*. Applied Research Center, Oakland, CA.

Magdoff, F. and Foster, J.B. (2011) *What Every Environmentalist Needs to Know about Capitalism*. Monthly Review Press, New York.

McAdam, D., McCarthy, J.D. et al. (1996) *Comparative Perspectives on Social Movements*. Cambridge University Press, Cambridge.

Mcgarvey, D. (2017) *Poverty Safari*. Picador, London.

Meyer, A. (2000) *Contraction and Convergence: The Global Solution to Climate Change*. Green Books, Cambridge.

Monbiot, G. (2020) 'Electric cars won't solve our pollution problems – Britain needs a total transport rethink', *The Guardian,* 23 September.

Monbiot, G. (2017a) 'How Labour could lead the global economy out of the 20th century', *The Guardian*, 11 October.

Monbiot. G. (2017b) 'A lesson from Hurricane Irma: capitalism can't save the planet – it can only destroy it', *The Guardian*, 13 September.

Morena, E. and Krause, D. (2018) (eds.) *Mapping Just Transition(s) to a Low-Carbon World: A Report of the Just Transition Research Collaborative.* United Nations Research Institute for Social Development (UNRISD), Geneva.

Oishi, S., Kesebir, S. and Diener, E. (2011) 'Income inequality and happiness', *Psychological Science,* vol 22, no 9, pp. 1095–1100.

Oxfam (2013) 'The cost of inequality: how wealth and income extremes hurt us all', www.oxfam.org/sites/www.oxfam.org/files/cost-of-inequality-oxfam-mb180113.pdf.

Oxfam (2016) 'An economy for the 1%: How privilege and power in the economy drive extreme inequality and how this can be stopped', http://policy-practice.oxfam.org.uk/publications/an-economy-for-the-1-how-privilege-and-power-in-the-economy-drive-extreme-inequ-592643.

Pandit, R. and Laband, D.N. (2009) 'Economic well-being, the distribution of income and species imperilment', *Biodiversity and Conservation,* vol 18, pp. 3219–3233.

Peoples Summit (2012) 'Final declaration of the People's Summit in Rio +20', 29 June. http://rio20.net/en/propuestas/final-declaration-of-the-peoples-summit-in-rio-20/.

Peterson, A. and Lupton, D. (1996) *The New Public Health: Health and the Self in the Age of Risk.* Sage Publications, Thousand Oaks, CA.

Piff, P.K., Stancato, D.M., Côté, S., Mendoza-Denton, R. and Keltner, D. (2012) 'Higher social class predicts increased unethical behavior', *Proceedings of the National Academy of Sciences of the United States of America,* vol 109, no 11, pp. 4086–4091.

Raworth, K. (2017) *Doughnut Economics: Seven Ways to Think Like a 21st-Century Economist.* Chelsea Green Publishing, Vermont.

Rehm, J. and Shield, K.D. (2019) 'Global burden of disease and the impact of mental and addictive disorders', *Current Psychiatry Reports,* vol 21 no 2, p. 10.

Rosenberg, A. (2017) 'Strengthening just transition policies in international climate governance', Policy Analysis Brief, The Stanley Foundation, Muscatine, IA.

Royal College of Psychiatrists (2010) 'No health without public mental health', Royal College of Psychiatrists Position statement PS4/2010, RCP, London.

Sanders, B. (2009) *The Green Zone: The Environmental Costs of Militarism.* AK Press, Chico, CA.

Schlosberg, D. (2003) 'The justice of environmental justice: Reconciling equity, recognition, and participation in a political model', in A. Light and A. de-Shalit (eds.), *Moral and Political Reasoning in Environmental Practice.* MIT Press, Cambridge, MA.

Schlosberg, D. (2004) 'Reconceiving environmental justice: Global movements and political theories', *Environmental Politics,* vol 13, no 3, pp. 517–540.

Schlosberg, D. (2007) *Defining Environmental Justice: Theories, Movement and Nature.* Oxford University Press, Oxford.

Seligman, M.E.P. (1972) 'Learned helplessness', *Annual Review of Medicine,* vol 23, no 1, pp. 407–412.

Social Mobility Commission (2017) *Social Mobility Policies between 1997 and 2017: Time for Change,* SMC, London.

Spivak, G.C. (1988) 'Can the subaltern speak?' in C. Nelson and L. Grossberg (eds.), *Marxism and the Interpretation of Culture.* University of Illinois Press, Champaign, IL.

Steffen, W., Richardson, K., Rockström, J., Cornell, S.E., Fetzer, I., Bennett, E.M. et al. (2015) 'Planetary boundaries: Guiding human development on a changing planet', *Science,* vol 347, no 6223, 1259855.

Stevis, D., Kraus, D. and Morena, E. (2020) 'Introduction: The genealogy and contemporary politics of just transitions', in E. Morena, D. Krause and D. Stevis (eds.), *Just Transitions: Social Justice in the Shift Towards a Low-Carbon World.* Pluto Press, London.

Torras, M. and Boyce, J.K. (1998) Income, inequality, and pollution: a reassessment of the environmental Kuznets Curve, *Ecological Economics,* vol. 25, no 2, pp. 147-160

Torras, M., Moskalev, S.A., Hazy, J.K. and Ashley, A.S. (2011) 'An econometric analysis of ecological footprint determinants: implications for sustainability', *International Journal of Sustainable Society*, vol 3, pp. 258–275.

Veblen, T. 1994 [1899] *The Theory of the Leisure Class: An Economic Study in the Evolution of Institutions.* The Macmillan Company, New York.

Velicu, I and Barca, S. (2020) The Just Transition and its work of inequality, *Sustainability: Science, Practice and Policy*, vol 16, no 1, pp. 263-273.

Vona, V. and Patriarca, F. (2011) 'Income inequality and the development of environmental technologies', *Ecological Economics*, vol 70, pp. 2201–2213.

Wakefield, S. and Poland, B. (2005) 'Family, friend or foe? Critical reflections on the relevance and role of social capital in health promotion and community development', *Social Science and Medicine*, vol 60, no 12, pp. 2819–2832.

Weir, K. (2012) 'The pain of social rejection', *American Psychological Association*, vol 43, no 4, pp. 50–60.

Wilkinson, R. and Pickett, K. (2010) *The Spirit Level: Why Greater Equality Makes Societies Stronger.* Bloomsbury Publishing, New York.

Wilkinson, R. and Pickett, K. (2019) *The Inner Level: How More Equal Societies Reduce Stress, Restore Sanity and Improve Everyone's Well-Being.* Penguin Press, London.

INDEX

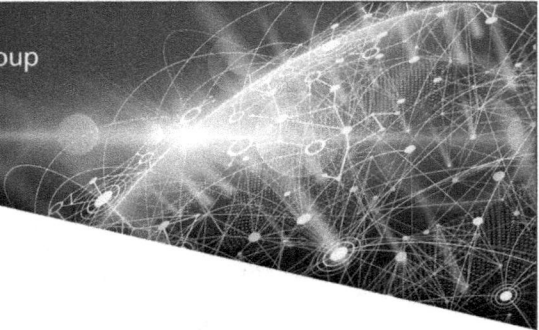